Weed Management

Weed Management

Reuben Kingsley
Editor

KOROS PRESS LIMITED
London, UK

Weed Management

© 2012

Printed in 2017 for Sale in the Indian Subcontinent

Published by
Koros Press Limited
3 The Pines, Rubery B45 9FF, Rednal,
Birmingham, United Kingdom

Tel.: +44-7826-930152
Email: info@korospress.com
www.korospress.com

ISBN: 978-1-78163-122-5

Editor: Reuben Kingsley

Printed in UK

British Library Cataloguing in Publication Data
A CIP record for this book is available from the British Library

10 9 8 7 6 5 4 3 2 1

Exclusively distributed by CBS Publishers & Distributors Pvt. Ltd.
Sales & Distribution Rights only for India, Pakistan, Bangladesh, Sri Lanka, Nepal and Bhutan.This book is not to be sold outside these territories.

Contents

Preface

The development of biodynamic agriculture began in 1924 with a series of eight lectures on agriculture given by philosopher Rudolf Steiner at Schloss Koberwitz in Silesia, Germany. The course was held in response to a request by farmers who noticed degraded soil conditions and a deterioration in the health and quality of crops and livestock resulting from the use of chemical fertilizers. At the course of eight lectures there were 111 attendees coming from six countries. An agricultural research group was subsequently formed to test the effects of biodynamic methods on the life and health of soil, plants andanimals. Today biodynamics is practiced in more than 50 countries worldwide. Demeter International is the primary certification agency for farms and gardens using the methods.

Biodynamic agriculturalists conceive of the farm as an organically self-contained entity with its own individuality, within which organisms are interdependent. "Emphasis is placed on the integration of crops and livestock, recycling of nutrients, maintenance of soil, and the health and well being of crops and animals; the farmer too is part of the whole." Cover crops, green manures and crop rotations are used extensively and the farms foster biodiversity. Biodynamic farms often have a cultural component and encourage local community. Some biodynamic farms use the Community Supported Agriculture model, which has connections with social threefolding. Steiner prescribed nine different preparations to aid fertilization which are the cornerstone of biodynamic agriculture, and described how these were to be prepared. Steiner believed that these preparations transferred supernatural terrestrial and cosmic "forces" into the soil. The prepared substances are numbered 500 through 508, where the first two are used for preparing fields whereas the latter seven are used for making compost. A long term trial evaluating the biodynamic farming system in comparison with organic and conventional farming systems, found that preparations have influence on soil structure and micro-organisms enhancing soil fertility and increasing biodiversity. The approach considers that there are astronomical influences on soil and plant development, specifying, for example, what phase of the moon is most appropriate for planting, cultivating or harvesting various kinds of crops. This aspect of biodynamics has been termed "astrological" in nature. Biodynamic agriculture has focused on open pollination of seeds (permitting farmers to grow their own seed) and the development of locally adapted varieties. The seed stock is not controlled by large, multinational seed companies.

The book should prove helpful in not only teaching the weed science courses but also in revamping and development of need-based course curriculum.

—Editor

Chapter 1

Cannabis in Australia

Cannabis is the most widely used illicit drug in Australia, with a reported one-third of all Australians aged 14 or older (33.5%, about 5.8 million) having tried cannabis and 1.6 million using it in the past year. Professor and Director of the National Cannabis Prevention and Information Centre (NCPIC) Jan Copeland estimates that 750,000 Australians use cannabis every week, and approximately 300,000 smoke it on a daily basis. Australia has one of the highest cannabis prevalence rates in the world, and research shows that Australia'sindigenous population has markedly higher levels of cannabis use. Cannabis is illegal in Australia; however, unlike the USA the country has largely avoided a punitive drug policy. Instead, it has developed harm-minimisation strategies and a treatment framework embedded in a law-enforcement regime.

History

Early History

The first hemp seeds were brought to Australia with the First Fleet at the request of Sir Joseph Banks, who marked the cargo "for commerce" in the hope that hemp would be produced commercially in the new colony. For 150 years early governments in Australia actively supported the growing of hemp with gifts of land and other grants, and the consumption of cannabis in Australia in the 19th century was believed to be widespread. It was popular as a medicine, and was used as an intoxicant by members of the literati; Marcus Clarke, author of the great Australian novel *For the Term of his Natural Life*, experimented with cannabis as an aid to writing. A short story he wrote, *Cannabis Indica*, was written under the influence of cannabis; members of Melbourne's bohemian Yorrick Club (of which Clarke was a member) were notorious cannabis users. Until the late 19th century, "Cigares De Joy" (cannabis cigarettes) were widely available; these claimed to "give immediate relief in cases of asthma, cough, bronchitis, hay-fever, influenza and shortness of breath". Like many developed nations Australia first responded to the issue of cannabis use in the 1920s, acting as a signatory to the 1925 Geneva Convention on Opium and Other Drugs that saw the use of cannabis restricted for medicinal and scientific purposes only. Cannabis was grouped with morphine, cocaine and heroin, despite cannabis' rare use as a medicine or remedy in Australia at the time. This prohibition model was applied with little research into cannabis use in Australia. Most drug-related laws enacted by jurisdictions of Australia during this time were related to opium but as a result of pressure from the United Kingdom, Australia began implementing local laws consistent with the Geneva Convention. According to McDonald and others, in 1928 the state of Victoria enacted legislation that prohibited the use of cannabis; other states followed suit slowly over the next three decades.

As in other Western countries, cannabis use was perceived as a significant social problem in Australia; new drug control laws were enacted at the state and federal level, and penalties for drug offences were increased. In 1938, cannabis was outlawed in Australia as a result of a *Reefer Madness*-style shock campaign; the newspaper *Smith's Weekly* carried a headline reading "New Drug that Maddens Victims". This campaign introduced the word "marijuana" to Australia, describing it both as "an evil sex drug that causes its victims to behave like raving sex maniacs" and "the dreaded sex drug marijuana". The campaign was only moderately successful; it instilled in a generation the negative effects of the drug and its impact on society, but did not stop an increase in demand and usage.

The Sixties

The 1960s saw an increase in the use of cannabis, heroin and LSD as part of political and social opposition to the Vietnam War, and this resulted in most Australian states gradually moving to a prohibitionist and criminal-justice orientation. Right-wing Australian politicians like Queensland premier Joh Bjelke-Petersen and NSW premier Robert Askin supported Nixon's War on Drugs in America, calling for a crackdown on Australian youth culture. Following the fall of the Whitlam government in 1975, these politicians launched a Nixon-style war on drugs in Australia.

In the late 1960s organised drug trafficking developed in Sydney with the arrival of US servicemen on leave from the Vietnam War, and the local drug markets expanded to meet their requirements. The 1970s were considered the first "decade of drugs", marked by the public's growing financial capacity to support drug use and an increase in young people affected by unemployment. As a result, the 1970s were also the decade of Royal Commissions and inquiries to deal with the "drug problem". In 1964, with the discovery of hundreds of acres of wild hemp growing in the Hunter Valley in NSW, authorities responded with a massive eradication campaign. However, the baby-boomers of the 60s responded to the "evil threat" in a very different manner to the previous generation, with groups of surfers and hippies flocking to the Hunter Region in search of the wild weed which was described in reports as "a powerful psychoactive aphrodisiac". These groups became known as the Weed Raiders—legendary characters, bearing tales of plants up to three metres tall.

1970s to Present

In 1973 tribes of hippies, championing the cause of decriminalising cannabis, attended the Aquarius Festival in the Northern NSW town of Nimbin. When police tried to arrest revelers who were openly smoking marijuana, the crowd of 6,000 rioted. This is believed to be the beginning of Nimbin's reign as the epicentre of cannabis culture and social counterculture in Australia. Nimbin is home to the Hemp Embassy, founded by activist pioneer Michael Balderstone, and the MardiGrass, an annual festival dedicated to cannabis which began in 1993.

According to Jiggens, by 1977 there was again talk of decriminalisation of cannabis in New South Wales, following the decriminalisation of cannabis in nine US states. The Joint Committee upon Drugs of the NSW Parliament recommended the removal of jail sentences for personal use of cannabis, and NSW Premier Neville Wran outlined a plan to remove jail sentences for people convicted of possessing cannabis for personal use. He said that cannabis use was widespread and that "tens of thousands of parents whose sons and daughters smoke marijuana" would not want their children to carry "the stigma of being a jailed, convicted criminal". However, the disappearance of local political and community leader Donald Mackay in Griffith, NSW in July 1977 placed the issue of the nexus between illicit drug production, organised crime and police corruption before the public; this was due to Mackay's revelations about large-scale marijuana growing in the Riverina area. His inquiries led to the largest cannabis seizure in Australian history at Coleambally, 60 km south of Griffith, in November 1975. The plantation spread over 31 acres (130,000 m^2) and was estimated to be capable of producing 60 tonnes of cannabis. The NSW Royal Commission into Drug Trafficking (the Woodward Inquiry) was sparked by Mackay's disappearance, and the story was brought to life as an acclaimed television miniseries *Underbelly: A Tale of Two Cities.*

Things were different in Queensland. In August 1976 NSW Police conducted a predawn raid on the Tuntable Falls Co-operative, located just south of Nimbin; a few weeks later the Cedar Bay commune, located in far northern Queensland, was raided by Queensland Police. "Using a helicopter, a naval patrol boat and four wheel drives they rounded up the members of the isolated community. Finding only a small quantity of marijuana, the police discharged firearms into water tanks and burned down the hippie's houses before they left". Joh Bjelke-Petersen defended the police action (including the burning of houses on the commune), declaring he was "tough on drugs". His accomplice in the Cedar Bay raid was the young John Howard (then Minister for Business), who continued Bjelke-Petersen's politics during his time as Prime Minister from 1996 to 2007. This would develop into an international news story.

In terms of the broader population, cannabis was not widely used in Australia until the 1970s. Legislation reflected increased usage of cannabis; in 1985 the National Campaign Against Drug Abuse was introduced, which was an assessment of illicit drug use among the general population. Prior to 1985, it was concluded that cannabis use amongst Australians rose from the early 1970s throughout the 1980s. Donnelly and Hall report that in a survey conducted in 1973, 22% of Australians aged 20–29 years reported ever using cannabis. This rose to 56% in 1985, and school surveys show a marked increase in cannabis use during the 1970s and 1980s. The rise in the use of cannabis continued into the 1990s with the 1998 household survey recording the highest prevalence of cannabis use, with 39% of those surveyed using cannabis at least once and 18% reporting cannabis use in the past year. By 2001 the lifetime rate had fallen to one-third of the population, where it currently remains.

According to Donnelly and Hall, although changes in willingness to divulge illicit drug use and changing survey protocol and design are likely to have contributed to the change in prevalence, the extent and consistency of the increase suggests that an actual rise in cannabis use has occurred.

Usage

According to Copeland and others, cannabis in Australia is commonly smoked as a cluster (or "cone") of the flowering heads or leaves of the female plant. Less commonly, the plant resin is smoked;

this is known as hash or hash oil. Typically, cannabis is smoked using a bong or joint and is often mixed with tobacco. Occasionally, cannabis is eaten or brewed as tea.

Cannabis use varies with age, and is most prevalent among Australians in their 20s and 30s. Patterns of use are similar to those throughout the developed world with heaviest use occurring in the early 20s, followed by a steady decline into the 30s. 90% of experimental or social recreational users of cannabis do not go on to use the substance daily or for a prolonged period; most discontinue its use by their late 20s.

According to the 2007 National Drug Strategy Household Survey, cannabis was used at least once by one-third of all Australians aged 14 years or older, and 1.6 million people reported using cannabis in the preceding 12 months. Of 12- to 15-year-olds 2.7% reported using cannabis in the previous 12 months, compared with 15% of 16- and 17-year-olds and 19% of 18- and 19-year-olds.

Results indicate that males aged 14 years or older were more likely than their female counterparts to have ever used cannabis (37.1% versus 30.0%), and one in five teenagers aged 14 to 19 reported having used cannabis. This gender differential is seen across all age groups except the 14- to 19-year-olds, in which there is little difference between males and females in terms of lifetime and past-year use.

Of the entire population, those aged 30 to 39 years were the most likely (54.6%) to have used cannabis at some time in their lives. According to McLaren and Mattick, the lower proportion of cannabis use among older age groups compared with younger users is even more striking when recent use is assessed; males aged 14 and older were more likely than corresponding females to have used cannabis in the previous 12 months (1.0 million and 0.6 million, respectively). 12.9% of teenagers aged 14 to 19 had used cannabis in the previous 12 months; those aged 20 to 29 were the most likely age group to have used cannabis in the previous 12 months, with one in five having done so. According to Hall although rates of cannabis use are considerable, most people who use cannabis do so infrequently. According to the 2004 household survey, approximately half of all recent cannabis users used the drug less than once a month. However, the proportion of recent cannabis users who use cannabis every day is not considered trivial; it is cited at 16% by the Australian Institute of Health and Welfare. Those aged 30 to 39 were most likely to use cannabis every day. The 2004 household survey also shows that of all respondents who used cannabis on a regular basis, the average number of cones or joints smoked on any one day was 3.2. Statistics show that between 1995 and 2007 (after peaking in 1998), the proportion of both males and females aged 14 years or older who had used cannabis in the previous 12 months declined steadily. Between 2004 and 2007, the decline was significant. Recent cannabis use dropped steadily since 1998 and significantly between 2004 and 2007—from 11.3% to 9.1%, the lowest proportion seen since 1993. Cross-sectional analysis of household survey data shows the age of initiation into cannabis decreasing over time. According to the Mental Health Council of Australia in 2006, the average age of first use for 12– to 19-year-olds was 14.9 years—significantly lower than in previous years.

Cannabis is considered relatively easy to obtain in Australia, with 17.1% of the population recording that they were offered (or had the opportunity to use) cannabis.

Indigenous Australia

Statistics show that Aboriginal and Torres Strait Islander communities have a lower life expectancy and higher rates of disease and injury than the non-indigenous population. Well-documented historical and social factors have contributed to the widespread use of tobacco and alcohol among indigenous communities and according to Perkins, Clough and others, the use of illicit drugs (cannabis in particular) is higher among Aboriginal and Torres Strait Islander peoples than among the non-indigenous population of Australia.

Little detailed information is available on cannabis use in urban or remote indigenous communities. Copeland and others cite 2001 National Drug Strategy Household Survey results showing that 27% of Aboriginal and Torres Strait Islander respondents reported using cannabis in the last 12 months, compared with 13% of non-indigenous Australians. However, these results are likely to under-report cannabis use in non-urban Aboriginal populations; communities are often small, isolated and highly mobile, making data collection problematic. What little detailed information is available on remote indigenous communities comes mainly from targeted studies of several communities in the Top End of Australia's Northern Territory.

Studies that do provide information on cannabis use within the indigenous population show pattern of problematic cannabis abuse that exceeds that seen in the mainstream non-indigenous population. A survey conducted in the mid-1980s by Watson and others failed to detect any cannabis use in Top End indigenous communities. However, by the late 1990s the Aboriginal Research Council provided information suggesting that cannabis was used by 31% of males and 8% of females in eastern Arnhem Land. A further study in 2002 found that cannabis was being used regularly by 67% of males and 22% of females aged 13 to 36.

As part of the 2004 National Drug Strategy, a survey was conducted assessing drug use among indigenous populations living in urban areas. Results showed that 48% had tried cannabis at least once, and 22% had used cannabis in the previous year. Regular cannabis use (at least weekly) was also more common among Aboriginal and Torres Strait Island communities than non-indigenous groups (11% and 4%, respectively).

A state-wide survey of students in New South Wales (NSW) indicated that the use of cannabis is significantly higher among indigenous students. Researchers noted that, after adjusting for socio-demographic variables, indigenous students were 1.6 times more likely to have ever tried cannabis than non-indigenous students.

The data describing cannabis use in the indigenous population compared with non-indigenous use varies in the ratio of recent cannabis use to those respondents who have ever used cannabis. In the non-indigenous population, rates of cannabis use in the last 12 months are a third of those ever using cannabis; however, researchers found only a few percentage-points' difference between rates of regular and lifetime use within the indigenous population.

According to McLaren and Mattick, the reasons for high rates of cannabis use among Aboriginal and Torres Strait Islander communities are complex and likely to be related to the social determinants of drug use. Risk factors associated with harmful substance use are often related to poor health and social well-being, stemming from the alienation and dispossession experienced by this population. Spooner and Hetherington confirm that many of the social determinants of harmful substance abuse are disproportionately present in Aboriginal and Torres Strait Islander communities.

Legislation and Policy

In 1913 Australia signed the International Hague Convention on Narcotics, and extended importation controls over drugs other than opium. 1921 saw the first international drug treaty (the Opium Convention), and in 1925 the Geneva Convention on Opium and Other Drugs saw restrictions imposed on the manufacture, importation, sale, distribution, exportation and use of cannabis, opium, cocaine, morphine and heroin for medical and scientific purposes only. In 1926 the Commonwealth Government banned the importation of cannabis;

in 1928 Victoria passed the Poisons Act and became the first state to control cannabis, followed by South Australia (1934), NSW (1935), Queensland (1937), Western Australia (1950) and Tasmania (1959). In 1940 the Commonwealth extended import restrictions on Indian hemp, including preparations containing hemp.

Proposed Decriminalisation

The 1978 NSW Joint Parliamentary Committee Upon Drugs supported the decriminalisation of cannabis; under the proposal, personal use of cannabis would no longer be an offence and users would be given bonds and probation. Trafficking in cannabis would carry severe penalties. However, the 1979 Australian Royal Commission of Inquiry into Drugs recommended against decriminalisation, concluding that such a step would contravene the UN Single Convention on Narcotic Drugs and lead to calls for the decriminalisation of other drugs. The recommendation was that the consideration of decriminalisation be delayed for another 10 years.

In 1985, against a backdrop of growing awareness at community and government levels of illicit drug use at a national level, the National Campaign Against Drug Abuse (NCADA) was established.

Since 1985, the national drug policy in Australia has been based on the principle of harm minimalisation; the National Campaign against Drug Abuse has since become the National Drug Strategy. The National Cannabis Strategy 2006–2009 was endorsed in 2006.

Legal Consequences

Unlike the United States, Australia has largely avoided a punitive drug policy, developing instead harm-minimisation strategies and a treatment framework embedded in a law-enforcement regime. Import and export of cannabis is illegal, and federal penalties apply. Offences can lead to sentences of up to life imprisonment for cases involving import or export of commercial quantities (100 kg and above for cannabis, 50 kg and above for cannabis resin and 2 kg and above for cannabinoids). Offences for quantities below a commercial quantity have lesser penalties attached. Federal offences also target the commercial cultivation of cannabis, domestic trafficking and possession. However, most cannabis offences committed are dealt with under state and territory legislation.

According to the Ministerial Council on Drug Strategy, the National Drug Strategy and its substance-specific strategies were written for the general population of Australia. The Aboriginal and Torres Strait Islander Peoples Complementary Action Plan 2003–2006 was developed as a supplement to the national action plans so that these plans could be applied to Australia's indigenous communities.

At a national level, there is no overriding law that deals with cannabis-related offences; instead, each state and territory enacts its own legislation. According to Copeland and others, while some jurisdictions enforce criminal penalties for possession, use and supply, others enact civil penalties for minor cannabis offences. Conviction for a criminal offence will attract a criminal record and can be punishable by jail time and harsh fines. Civil penalties, however, do not result in a criminal record and are generally handled by lesser fines, mandatory treatment and diversion programmes. In fact, all Australian states and territories have implemented systems where non-violent, minor and early cannabis offenders are diverted from the legal system. Although violent offenders and dealers are excluded, cannabis-cautioning schemes have been implemented in several states. Offenders are issued a caution notice rather than facing criminal proceedings; cautioning systems include an educational component on the harm of cannabis. Some also contain mandatory counselling or more substantial treatment for repeat offenders.

National Cannabis Prevention and Information Centre (NCPIC) details the penalty systems in place throughout Australia; in the Australian Capital Territory (ACT), a civil-penalty system for possession of small amounts of cannabis was introduced in 1993. Possession of up to 25 g or two non-hydroponic plants attracts a fine of 100 Australian dollars, due within 60 days. Offenders can choose to attend the Alcohol and Drug Program (ACT Alcohol, Tobacco and Other Drugs Strategy). In South Australia possession of small quantities of cannabis is decriminalised, attracting a fine similar to that for a parking ticket. However, penalties for growing cannabis have become harsher since the advent of widespread large-scale cultivation. There is much confusion on the subject, with many believing that possession of a small amount of cannabis is legal. In Western Australia, possession of up to 30 g, two non-hydroponic plants, or smoking equipment attracts a fine of up to A$200, with an option to attend a cannabis education session. (Drug Aware) Any amount exceeding this is dealt with by the criminal court. The Northern Territory has a similar civil penalty to that in Western Australia.

In New South Wales, Queensland, Victoria and Tasmania, possession and use of cannabis is a criminal offence; however, it is unlikely that anyone caught with a small amount will be convicted. Diversion programs in these states aim to divert offenders into education, assessment and treatment programs. In New South Wales, if one is caught with up to 15 g of cannabis, at police discretion up to two cautions can be issued. In Tasmania up to three cautions can be issued for possession of up to 50 g of cannabis, with a hierarchy of intervention and referrals for treatment with each caution. Similarly, in Victoria up to 50 g of cannabis will attract a caution and the opportunity to attend an education program (Victoria Cannabis Cautioning Program); only two cautions will be issued. In Queensland, possession of cannabis or any schedule 1 or 2 drug specified in the Drugs Misuse Regulation 1987 carries a maximum prison sentence of 15 years; however, jail terms for minor possession is very rare. Possession of smoking paraphernalia is also a criminal offence in Queensland. However, under the Police Powers and Responsibilities Act 2000 a person who admits to carrying under 50 g (and is not committing any other offence) must be offered a drug diversion program.

With the rapid expansion in hydroponic cannabis cultivation, the Australian Drug Misuse and Trafficking Act (1985) was amended in 2006; the amount of cannabis grown indoors under hydroponic conditions that qualifies as a "commercial quantity" or as a "large quantity" was reduced.

Medicinal Use

The use of cannabis for any purpose is illegal in Australia. Clinical trials of cannabis for medicinal purposes have been suggested; however, no jurisdiction has indicated it will conduct trials in the near future and there does not appear to be widespread support for medical trials at the government level. Support for a change in legislation permitting the use of marijuana for medical purposes remained relatively unchanged between 2004 and 2007. Two-thirds (68.6%) of respondents in the 2007 NDSHS survey supported "a change in legislation permitting the use of marijuana for medical purposes" and almost three-quarters (73.6%) supported "a clinical trial for people to use marijuana to treat medical conditions". Females were slightly more likely than males to support either of these measures.

Supply

Statistics on the prevalence of cannabis use indicate the existence of high demand for the drug. As cannabis is an illicit drug, a sizable black market exists to meet demand.

Prevalence and Price

The prevalence of cannabis in Australia indicates that the drug is widely available. The University of New South Wales' National Drug and Alcohol Research Centre's *Drug Trends Bulletin* for October

2009 shows that 58% of cannabis users in NSW believe hydroponically-grown cannabis to be "very easily" available; 43% believe bush-grown cannabis is "very easy" to find. 0% considered hydro cannabis "very difficult" to find and 5% considered bush-grown cannabis to be "very difficult" to find. The results show that figures for the ACT are lower (42% believe hydroponically-grown cannabis is "very easy" to find, as do 29% for bush-grown cannabis. 3% and 7%, respectively, believe that cannabis is "very difficult" to find). Victoria shows similar figures to NSW; 66% and 32%, respectively, believe cannabis is "very easy" to find and 0% and 3%, respectively, believe it is "very difficult" to find). Tasmania shows similar statistics. In South Australia fewer people consider cannabis (either hydroponically- or bush-grown) "very easy" to find (32% and 37% respectively), with the majority considering it "easy" to find (46% and 21%). Western Australia reports similar statistics as South Australia, as does the Northern Territory. Queensland reports statistics similar to NSW with 64% and 56% of respondents reporting hydroponically grown cannabis and bush cannabis, respectively, "very easy" to find and 3% and 6%, respectively, considering it "very difficult" to find. The majority of cannabis is domestically produced, with outdoor and hydroponic cultivation common in all states and territories. Single and others note that Australia's climate and the amount of space available is conducive to outdoor cultivation. According to the Australian Crime Commission (ACC) the average price for one gram of cannabis ranged from A$20–A$35, although prices in remote areas can be significantly higher. In remote regions of the Northern Territories, for example, the price can reach $50–$100 for a gram. According to Stafford and Burns, an ounce of hydroponically grown cannabis has risen from A$300–$320 between 2008 and 2009; an ounce of bush weed has increased from A$200–$229. NDSHS notes that one in six Australians reported that they were offered or had the opportunity to use cannabis. The ACC reports that hydroponically-grown cannabis is described by 75% of the 2007 NDSHS respondents as being "easy" or "very easy" to obtain; "bush cannabis" (outdoor-grown cannabis), by contrast, is not as readily available and was reported by over half of the respondents as being "easy" to obtain.

Respondents in the National Drug and Alcohol Research Centre (NDARC) October 2009 *Drug Trends Bulletin* were asked to rate the purity and potency of cannabis. Statistics show that, in general, hydroponically-grown cannabis is considered to have high purity and potency (NSW 61%; ACT 54%; Victoria 58%; Tasmania 66%; South Australia 65%; Western Australia 69%; Northern Territory 38% [14%

low; 31% medium; 17% fluctuates]; Queensland 58%). Bush-grown cannabis is considered to have medium purity and potency (explained by the greater variables in production), with a number of respondents categorising bush grown cannabis as poor-quality. Respondents reported daily or near-daily use of cannabis. According to the 2007 NDSHS, 68.5% of cannabis users obtained cannabis from a friend or acquaintance. 4.8% acquired it from a relative, and 19.5% obtained it from a dealer. 7.2% claimed to have acquired the drug in another way, including "grew/made/picked it myself.

Seizures and Arrests

According to the Australian Crime Commission (ACC), cannabis accounted for the greatest proportion of national illicit drug arrests and seizures in 2007/2008 – 5409 kg (5,409,000 grams) were seized nationally over 12 months, accounting for 64% of illicit drugs seized in Australia. This equates to 41,660 cannabis seizures, or 68% of all seizures. 2007/2008 saw 52,465 cannabis arrests, a 7% decrease from figures for 2006/2007. The majority of arrests continue to occur in Queensland. Despite a slight decrease from 2006, cannabis continues to be the most commonly detected drug amongst police detainees. Self-reporting within this group identifies hydroponically-grown heads as both the preferred and actual form of cannabis used by the majority of detainees.

Advocacy

Support for the legalisation of illicit drugs declined slightly between 2004 and 2007 and support for the legalisation for personal use of cannabis fell between 2004 and 2007, from 27.0% to 21.2%. Males were more likely than females to support legalisation (in 2007, 23.8% versus 18.5%). Most states and territories have policies or legislation in place which are designed to reduce the penalties for cannabis possession. The objective, according to the Australian Illicit Drug Guide, is to reduce backlogs in the judicial system caused by what are considered minor cannabis offences and to divert offenders into treatment and counselling. Cannabis-cautioning programs operate in Victoria, NSW, ACT, WA and Tasmania as part of the Illicit Drug Diversion Initiative. These programmes are policy-based (rather than legislation-based) approaches. Most states also have separate cautioning systems for juvenile offenders.

Cannabis Culture

The Nimbin Hemp Embassy is a non-profit association that was established in 1992. The Embassy's objectives are cannabis law reform via an education program for the community about hemp products and cannabis and "promoting a more tolerant and compassionate attitude to people in general". According to the HEMP Embassy website, "the Nearly NORML Nimbin group formed in 1988 as the district's first enduring drug law reform outfit and later became Nimbin HEMP- Help End Marijuana Prohibition- then later in 1992 the name changed to the Nimbin HEMP Embassy. Generally the group discussed the cannabis laws of NSW and how they might be changed" In 1993, as a passive response to police raids in Nimbin and increasingly negative local reaction, the HEMP Embassy created the inaugural "Let It Grow" May Day rally and street parade, a celebratory and non-provocative form of political action. This became the first-ever MardiGrass, now a well-known cannabis-law-reform rally and festival held annually in the town of Nimbin.

The Australian Help End Marijuana Prohibition (HEMP) political party has a number of objectives, including:

- endorsing candidates in federal elections
- legalising cannabis in all states and territories in Australia for
  - o personal use
  - o medical and therapeutic use
  - o industrial purposes
- collecting and disseminating knowledge relating to any or all of the Party's aims
- campaigning (and lobbying) in all sectors of the community
- organising fundraising for the Party
- conducting and facilitating research relating to any or all of the Party's aims
- applying for public funding for electoral purposes, in accordance with the provisions of the Australian Electoral Act (1918) as amended
- do all which may be necessary, expedient or desirable to carry out the aims of the Party.

The aims of the party centre on decriminalising both possession of small amounts of cannabis and the cultivation of cannabis for personal use, as well as legalising cannabis for medical use. A more radical proposal is that of "drug-free" zones which would "address issues of public consumption of cannabis through community policing" and the party supports greater funding for treatment services. In 2001 and 2004 the National President of the HEMP party, Michael

Balderstone, ran as a Senate candidate. The party did not contest the 2007 Federal elections because it had been de-registered and could not re-register in time. It is campaigning to enroll more members, to be eligible to register again.

Following in the footsteps of the Cannabis Cup in the Netherlands, the Cannabis Cup in Australia is a competition run by MardiGrass to judge strains of cannabis. Growers submit samples of their crop for judging and the Hemp Olympics, held at MardiGrass, includes events such as bong throwing, joint rolling and "a growers' Ironperson competition, which requires participants to crawl through lantana tunnels dragging large bags of fertilizer".

There is increasing interest in hemp in Australia—for example, the HEMP Bar in Nimbin, which champions the hempseed food revolution in Australia with income donated to the HEMP Party, and a recent case in the media detailing a hemp grower on the Northern Beaches of Sydney who has legally grown 500 plants in his backyard. *The Sydney Morning Herald* describes cultivator Richard Friar as a hemp evangelist—a firm believer in the world-changing potential of this most versatile of plants, which can be used in everything from food to fabrics and building materials. With permission from the NSW Department of Primary Industries, Friar and his wife are part of a pilot project aimed at educating farmers to the benefits of growing hemp for its byproducts from food to fabric. The author also notes that in December 2009, Friar applied to Food Standards Australia New Zealand for permission to sell the seed for human consumption; approval is expected. Wendy Friar is quoted that hemp is a superfood; "It's 23% protein, and has more Omega 3 and Omega 6 than virtually any other source, including fish. In the early 1800s, Australia was twice saved from famine by eating virtually nothing but hemp seed for protein and hemp leaves for roughage".

OZ Stoners is undoubtedly Australia's oldest and largest on-line cannabis community and cannabis information source. Boasting Australia's oldest cannabis forums it also offers comprehensive cannabis strain & seed bank information through its cannabis culture related guides and associated domains, as well as official weekly questions to the NCPIC directly from the on-line Australian cannabis community, on the hour cannabis news gathered from both Australian & international sources, legal forums & information, personal stories regarding the effects of cannabis prohibition.

Cannabis Culture Magazine is not available in Australia, but regularly comments on the cannabis culture of the country both in print and on-line. StickyPoint Magazine bills itself as "the Australian Cannabis Lifestyle and News Magazine", and is available both in print and online; however, due to state laws related to the R18+ classification of the magazine, *StickyPoint* is not available in Queensland.

The 420 Australia Cannabis, Marijuana and Hemp Portal bills itself as "an information site on the many potential industrial, medicinal and recreational applications of hemp, marijuana & cannabis seeds for Australia & New Zealand, and web platform for law (policy) reform and industrial development".

Medical Cannabis

Medical cannabis refers to the use of parts of the herb cannabis (also referred to as medical marijuana) as a physician-recommended form of medicine or herbal therapy, or to synthetic forms of specific cannabinoids such as THC as a physician-recommended form of medicine. The *Cannabis* plant from which the cannabis drug is derived has a long history of medicinal use, with evidence dating back to 2,737 BCE. Synthetic cannabinoids, such as Marinol and Cesamet, are available as prescription drugs in some countries. A number of studies, some disputed, claim that medical cannabis relieves symptoms and is helpful in the treatment of many diseases.

Use

The medicinal value of cannabis is controversial. A large majority of national governments do not recognise the use of plant parts from the plant Cannabis Sativa as something that doctors can recommend to their patients. A number of these governments, including the U.S. Federal Government, allow treatment with one or more specific synthetic cannabinoids for one or more disorders.

Supporters of medical cannabis argue that cannabis does have several well-documented beneficial effects. Among these are: the amelioration of nausea and vomiting, stimulation of hunger inchemotherapy and AIDS patients, lowered intraocular eye pressure (shown to be effective for treating glaucoma), as well as gastrointestinal illness. Its effectiveness as an analgesic has been suggested—and disputed—as well.

There are several methods for administration of dosage, including vapourising or smoking dried buds, drinking, or eating extracts, and taking capsules. The comparable efficacy of these methods was the subject of an investigative study conducted by the National Institutes of Health. Synthetic cannabinoids are available as prescription drugs in some countries. Examples are Marinol (The United States and

Canada) and Cesamet (Canada, Mexico, the United Kingdom, and the United States).

While utilising cannabis for recreational purposes is illegal in many parts of the world, many countries are beginning to entertain varying levels of decriminalisation for medical usage, including Canada, Austria, Germany, the Netherlands, Spain, Israel, Italy, Finland, and Portugal. In the United States, federal law outlaws all use of herb parts from Cannabis, while some states have approved use of herb parts from Cannabis as medical cannabis in conflict with federal law. The United States Supreme Court has ruled in *United States v. Oakland Cannabis Buyers' Coop* and *Gonzales v. Raich* that the federal government has a right to regulate and criminalise cannabis, even for medical purposes. A person can therefore be prosecuted for a cannabis-related crime even if it is medical cannabis that is legal according to the laws of this state.

Clinical Applications

A 2002 review of medical literature by Franjo Grotenhermen states that medical cannabis has established effects in the treatment of nausea, vomiting, premenstrual syndrome, unintentional weight loss, insomnia, and lack of appetite. Other "relatively well-confirmed" effects were in the treatment of "spasticity, painful conditions, especially neurogenic pain, movement disorders, asthma, and glaucoma".

Preliminary findings indicate that cannabis-based drugs could prove useful in treating inflammatory bowel disease, migraines, fibromyalgia, and related conditions.

Medical cannabis has also been found to relieve certain symptoms of multiple sclerosis and spinal cord injuries by exhibiting antispasmodic and muscle-relaxant properties as well as stimulating appetite.

Other studies state that cannabis or cannabinoids may be useful in treating alcohol abuse, amyotrophic lateral sclerosis, collagen-induced arthritis, asthma, atherosclerosis, bipolar disorder, colourectal cancer, HIV-Associated Sensory Neuropathy depression, dystonia, epilepsy, digestive diseases, gliomas, hepatitis C,Huntington's disease, leukemia, skin tumors, methicillin-resistant *Staphylococcus aureus* (MRSA), Parkinson's disease, pruritus, post-traumatic stress disorder (PTSD), psoriasis, sickle-cell disease, sleep apnea, and anorexia nervosa. Controlled research on treating Tourette syndrome with a synthetic version of tetrahydrocannabinol, (brand nameMarinol) (the main psychoactive chemical found in cannabis), showed the patients taking Marinol had a beneficial response without serious adverse effects; other studies have shown that cannabis "has no effects on tics and increases the individuals inner tension". Case reports found that marijuana helped reduce tics, but validation of these results requires longer, controlled studies on larger samples.

Recent Studies

Safety of Cannabis

According to an approved statement from the US Department of Justice in 1988, "Nearly all medicines have toxic, potentially lethal effects. But marijuana is not such a substance. There is no record in the extensive medical literature describing a proven, documented cannabis-induced fatality. In practical terms, marijuana cannot induce a lethal response as a result of drug-related toxicity."

From January 1997 to June 2005, the U.S. Food and Drug Administration (FDA) reported zero deaths caused by the primary use of marijuana. Through that time period, 279 deaths were reported where marijuana was a possible "concomitant" drug used in conjunction with other drugs at the time of death. In contrast, common FDA-approved drugs which are often prescribed in lieu of marijuana (such as anti-emetics and anti-psychotics), were the primary cause of 10,008 deaths.

Alzheimer's Disease

Research done by the Scripps Research Institute in California shows that the active ingredient in marijuana, THC, prevents the formation of deposits in the brain associated with Alzheimer's disease. THC was found to prevent an enzyme called acetylcholinesterase from accelerating the formation of "Alzheimer plaques" in the brain more effectively than commercially marketed drugs. THC is also more effective at blocking clumps of protein that can inhibit memory and cognition in Alzheimer's patients, as reported in Molecular Pharmaceutics.

Mental Disorders

There has been evidence that smoking marijuana can have a positive effect on disorders such as Schizophrenia, bipolar disorder, or depression. In patients with bipolar disorder subjects have been shown to actually become better after smoking marijuana increasing the rate at which these patients go from high to low. In the case of depression many users have reported that their moods have become better. Research done on lab rats and animals has shown that marijuana can act as an anti-depressant but in other studies done on humans this is not the case, actually pushing the subjects further into their depression. A study of 50,000 Swedish soldiers who had smoked at least once were twice as likely to develop schizophrenia as those who had not smoked. The study concluded that either smoking caused a higher rate of schizophrenia, or that schizophrenics were more likely to be drawn to marijuana.

A study by Keele University commissioned by the British government found that between 1996 and 2005 there had been significant reductions in the incidence and prevalence of schizophrenia. From 2000 onwards there were also significant reductions in the prevalence of psychoses.

The authors say this data is "not consistent with the hypothesis that increasing cannabis use in earlier decades is associated with increasing schizophrenia or psychoses from the mid-1990s onwards".

A 10 year study on 1923 individuals from the general population in Germany, aged 14–24, concluded that cannabis use is a risk factor for the development of incident psychotic symptoms. Continued cannabis use might increase the risk for psychotic disorder.

Lung Cancer and Chronic Obstructive Pulmonary Disease

The evidence to date is conflicting as to whether smoking cannabis increases the risk of developing lung cancer or chronic obstructive pulmonary disease (COPD) among people who do not smoke tobacco. In 2006 a study by Hashibe, Morgenstern, Cui, Tashkin, *et al.* suggested that smoking cannabis does not, by itself, increase the risk of lung cancer. Several subsequent studies have found results suggesting the reverse, however many of these were not completed with proper scientific controls and have subsequently been discredited. Many studies did report a strongly synergistic effect, however, between tobacco use and smoking cannabis such that tobacco smokers who also smoked cannabis dramatically increased their already very high risk of developing lung cancer or chronic obstructive pulmonary disease by as much as 300%. Some of these research results follow below:

- In 2006, Hashibe, Morgenstern, Cui, Tashkin, *et al.* presented the results from a study involving 2,240 subjects that showed non-tobacco users who smoked marijuana did not exhibit an increased incidence of lung cancer or head-and-neck malignancies. These results were supported even among very long-term, very heavy users of marijuana.

Tashkin, a pulmonologist who has studied marijuana for 30 years, said, "It's possible that tetrahydrocannabinol (THC) in marijuana smoke may encourage apoptosis, or programmed cell death, causing cells to die off before they have a chance to undergo malignant transformation". He further commented that "We hypothesized that there would be a positive association between marijuana use and lung cancer, and that the association would be more positive with heavier use. What we found instead was no association at all, and even a suggestion of some protective effect."

- A case-control study of lung cancer in adults 55 years of age and younger found that the risk of lung cancer increased 8% (95% confidence interval (CI) 2–15) for each joint-year of cannabis smoking, after adjustment for confounding variables including cigarette smoking, and 7% (95% CI 5–9) for each pack-year of cigarette smoking, after adjustment for confounding variables including cannabis smoking.

- A 2008 study by Hii, Tam, Thompson, and Naughton found that marijuana smoking leads to asymmetrical bullous disease, often in the setting of normal CXR and lung function. In subjects who smoke marijuana, these pathological changes occur at a younger age (approximately 20 years earlier) than in tobacco smokers.

- Researchers from the University of British Columbia presented a study at the American Thoracic Society 2007 International Conference showing that smoking marijuana and tobacco together more than tripled the risk of developing COPD over just smoking tobacco alone. Similar findings were released in April 2009 by the Vancouver Burden of Obstructive Lung Disease Research Group. The study reported that smoking both tobacco and marijuana synergistically increased the risk of respiratory symptoms and COPD. Smoking only marijuana, however, was not associated with an increased risk of respiratory symptoms of COPD. In a related commentary, pulmonary researcher Donald Tashkin wrote, "...we can be close to concluding that marijuana smoking by itself does not lead to COPD".

Breast Cancer

According to a 2007 study at the California Pacific Medical Centre Research Institute, cannabidiol (CBD) may stop breast cancer from spreading throughout the body. These researchers believe their discovery may provide a non-toxic alternative to chemotherapy while achieving the same results minus the painful and unpleasant side effects. The research team says that CBD works by blocking the activity of a gene called Id-1, which is believed to be responsible for a process called metastasis, which is the aggressive spread of cancer cells away from the original tumour site.

HIV/AIDS

Investigators at Columbia University published clinical trial data in 2007 showing that HIV/AIDS patients who inhaled cannabis four times daily experienced substantial increases in food intake with little evidence of discomfort and no impairment of cognitive performance. They concluded that smoked marijuana has a clear medical benefit in HIV-positive patients.

In another study in 2008, researchers at the University of California, San Diego School of Medicine found that marijuana significantly reduces HIV-related neuropathic pain when added to a patient's already-prescribed pain management regimen and may be an "effective option for pain relief" in those whose pain is not controlled with current medications. Mood disturbance, physical disability, and quality of life all improved significantly during study treatment. Despite management with opioids and other pain modifying therapies, neuropathic pain continues to reduce the quality of life and daily functioning in HIV-infected individuals. Cannabinoid receptors in the central and peripheral nervous systems have been shown to modulate pain perception. No serious adverse effects were reported, according to the study published by the American Academy of Neurology. A study examining the effectiveness of different drugs for HIV associated neuropathic pain found that smoked Cannabis was one of only three drugs that showed evidence of efficacy.

Brain Cancer

A study by Complutense University of Madrid found the chemicals in marijuana promotes the death of brain cancer cells by essentially helping them feed upon themselves in a process called autophagy. The research team discovered that cannabinoids such as THC had anticancer effects in mice with human brain cancer cells and in people with brain tumors. When mice with the human brain cancer cells received the THC, the tumor shrank. Using electron microscopes to analyse brain tissue taken both before and after a 26- to 30-day THC treatment regimen, the researchers found that THC eliminated cancer cells while leaving healthy cells intact. The patients did not have any toxic effects from the treatment; previous studies of THC for the treatment of cancer have also found the therapy to be well tolerated. However, the mechanisms which promote THC's tumour cell–killing action are unknown.

Opioid Dependence

Injections of THC eliminate dependence on opiates in stressed rats, according to a research team at the *Laboratory for Physiopathology of Diseases of the Central Nervous System* (France) in the journal *Neuropsychopharmacology*. Deprived of their mothers at birth, rats become hypersensitive to the rewarding effect of morphine and heroin (substances belonging to the opiate family), and rapidly become dependent. When these rats were administered THC, they no longer developed typical morphine-dependent behaviour. In the striatum, a region of the brain involved in drug dependence, the production of endogenous enkephalins was restored under THC, whereas it diminished in rats stressed from birth which had not received THC. Researchers believe the findings could lead to therapeutic alternatives to existing substitution treatments.

In humans, drug treatment subjects who use cannabis intermittently are found to be more likely to adhere to treatment for opioid dependence. Historically, similar findings were reported by Edward Birch, who, in 1889, reported success in treating opiate and chloral addiction with cannabis.

Controlling ALS Symptoms

Recent research has been conducted on if the use of marijuana could control some of the symptoms of ALS or Lou Gehrig Disease. A survey was conducted on 131 people who suffered from ALS. The survey asked if the subjects had used marijuana in the last 12 months to control some of their symptoms. The survey resulted in 13 people who had used the drug in some form to control symptoms. The survey results found that cannabis was moderately effective in reducing symptoms of appetite loss, depression, pain, spasticity, drooling and weakness and the longest relief reported was for depression. The pattern of symptom relief was consistent with those reported by people with other conditions, including MS (Amtmann et al. 2004).

Spasticity in Multiple Sclerosis

A review of six randomised controlled trials of a combination of THC and CBD extracts for the treatment of MS related muscle spasticity reported, "Although there was variation in the outcome measures reported in these studies, a trend of reduced spasticity in treated patients was noted." The authors postulated that "cannabinoids may provide neuroprotective and anti-inflammatory benefits in MS." A small study done on whether or not marijuana could be used to control tremors of MS patients was conducted. The study found that there was no noticeable difference of the tremors in the patients. Although there was no difference in the tremors the patients felt as if their symptoms had lessened and their quality of life had improved. The researchers concluded that the mood enhancing or cognitive effects that cannabis has on the brain could have given the patients the effect that their tremors were getting better.

Treatment of Inflammatory Skin Disease

The abundant distribution of cannabinoid receptors on skin nerve fibres and mast cells provides implications for an anti-inflammatory, anti-nociceptive action of cannabinoid receptor agonists and suggests their putatively broad therapeutic potential

Medicinal Compounds

Cannabis contains over 300 compounds. At least 66 of these are cannabinoids, which are the basis for medical and scientific use of cannabis. This presents the research problem of isolating the effect of specific compounds and taking account of the interaction of these compounds. Cannabinoids can serve as appetite stimulants, antiemetics, antispasmodics, and have some analgesic effects. Five important cannabinoids found in the cannabis plant are tetrahydrocannabinol, cannabidiol, cannabinol, β-caryophyllene, and cannabigerol.

Tetrahydrocannabinol

Tetrahydrocannabinol (THC) is the primary compound responsible for the psychoactive effects of cannabis. The compound is a mild analgesic, and cellular research has shown the compound has antioxidant activity. THC is believed to interact with parts of the brain normally controlled by the endogenous cannabinoid neurotransmitter, anandamide. Anandamide is believed to play a role in pain sensation, memory, and sleep.

Cannabidiol

Cannabidiol (CBD), is a major constituent of medical cannabis. CBD represents up to 40% of extracts of the medical cannabis plant. Cannabidiol has been shown to relieve convulsion, inflammation, anxiety, cough and congestion, nausea, and inhibits cancer cell growth. Recent studies have shown cannabidiol to be as effective as atypical antipsychotics in treating schizophrenia. Because cannabidiol relieves the aforementioned symptoms, cannabis strains with a high amount of CBD may benefit people with multiple sclerosis, frequent anxiety attacks and Tourette syndrome.

Cannabinol

Cannabinol (CBN) is a therapeutic cannabinoid found in *Cannabis sativa* and *Cannabis indica*. It is also produced as a metabolite, or a breakdown product, of tetrahydrocannabinol (THC). CBN acts as a weak agonist of the CB_1 and CB_2 receptors, with lower affinity in comparison to THC.

β-Caryophyllene

Part of the mechanism by which medical cannabis has been shown to reduce tissue inflammation is via the compound β-caryophyllene. A cannabinoid receptor called CB2 plays a vital part in reducing inflammation in humans and other animals. β-Caryophyllene has been shown to be a selective activator of the CB2 receptor. β-Caryophyllene is especially concentrated in cannabis essential oil, which contains about 12–35% β-caryophyllene.

Pharmacologic THC and THC Derivatives

In the USA, the FDA has approved several cannabinoids for use as medical therapies: dronabinol (Marinol) and nabilone. These medicines are taken orally.

These medications are usually used when first line treatments for nausea and vomiting associated with cancer chemotherapy fail to work. In extremely high doses and in rare cases "psychotomimetic" side effects are possible. The other commonly used antiemetic drugs are not associated with these side effects.

Canasol is a cannabis-based medication for glaucoma that relieves intraocular pressure symptoms associated with late-stage glaucoma.

It was created by an optomologist, Dr. Albert Lockhart and Dr. Manley E. West, and began distribution in 1987. As of 2003, it was still being distributed in the United Kingdom, several US states, and several Caribbean nations.

It is notable for being one of the first cannabis-containing pharmaceuticals to be developed for the modern pharmaceutical market and being one of the few such pharmaceuticals to have ever been legally marketed in the United States.

The prescription drug Sativex, an extract of cannabis administered as a sublingual spray, has been approved in Canada for the adjunctive treatment (use along side other medicines) of both multiple sclerosis and cancer related pain. Sativex has also been approved in the United Kingdom, New Zealand, and the Czech Republic, and is expected to gain approval in other European countries. William Notcutt is one of the chief researchers that has developed Sativex, and he has been working with GW and founder Geoffrey Guy since the company's inception in 1998. Notcutt states that the use of MS as the disease to study "had everything to do with politics."

Criticism

One of the major criticisms of cannabis as medicine is opposition to smoking as a method of consumption. However, smoking is no longer necessary due to the development of healthier methods. Today, medicinal marijuana patients can use vapourisers, where the essential marijuana compounds are extracted and inhaled. This is somewhat similar to steaming vegetables to avoid cancerous by-products that are produced at higher temperatures. In addition, edible marijuana, which is produced in various baked goods, is also available, and has demonstrated longer lasting effects.

The United States Food and Drug Administration (FDA) issued an advisory against *smoked* medical marijuana stating that, "marijuana has a high potential for abuse, has no currently accepted medical use in treatment in the United States, and has a lack of accepted safety for use under medical supervision. Furthermore, there is currently sound evidence that smoked marijuana is harmful." The National Institute on Drug Abuse NIDA state that "Marijuana itself is an unlikely medication candidate for several reasons: (1) it is an unpurified plant containing numerous chemicals with unknown health effects; (2) it is typically consumed by smoking further contributing to potential adverse effects; and (3) its cognitive impairing effects may limit its utility".

The Institute of Medicine, run by the United States National Academy of Sciences, conducted a comprehensive study in 1999 to assess the potential health benefits of cannabis and its constituent cannabinoids. The study concluded that smoking cannabis is not recommended for the treatment of any disease condition, but did conclude that nausea, appetite loss, pain and anxiety can all be mitigated by marijuana. While the study expressed reservations about smoked marijuana due to the health risks associated with smoking, the study team concluded that until another mode of ingestion was perfected that could provide the same relief as smoked marijuana, there was no alternative. In addition, the study pointed out the inherent difficulty in marketing a non-patentable herb. Pharmaceutical companies will not substantially profit unless there is a patent. For those reasons, the Institute of Medicine concluded that there is little future in smoked cannabis as a medically approved medication. The report also concluded that for certain patients, such as the terminally ill or those with debilitating symptoms, the long-term risks are not of great concern.

Marinol was less effective than the steroid megestrol in helping cancer patients regain lost appetites. A phase III study found no difference in effects of an oral cannabis extract or THC on appetite and quality of life (QOL) in patients with cancer-related anorexia-cachexia syndrome (CACS) to placebo.

"Citing the dangers of marijuana and the lack of clinical research supporting its medicinal value" the American Society of Addiction Medicine in March 2011 issued a white paper recommending a halt to using marijuana as a medicine in U.S. states where it has been declared legal.

Harm Reduction

The harm caused by smoking can be minimised or eliminated by the use of a vapouriser or ingesting the drug in an edible form. Vapourisers are devices that heat the active constituents to a temperature below the ignition point of the cannabis, so that their vapours can be inhaled. Combustion of plant material is avoided, thus preventing the formation of carcinogens such as polyaromatic hydrocarbons, benzene and carbon monoxide. A pilot study led by Donald Abrams of UC San Francisco showed that vapourisers eliminate the release of irritants and toxic compounds, while delivering equivalent amounts of THC into the bloodstream.

In order to kill microorganisms, especially the molds *A. fumigatus*, *A. flavus* and *A. niger*, Levitz and Diamond suggested baking marijuana at 150 °C (302 °F) for five minutes. They also found that tetrahydrocannabinol (THC) was not degraded by this process.

Organisational Positions

A number of medical organisations have endorsed reclassification of marijuana to allow for further study. These include, but are not limited to:

- The American Medical Association
- The American College of Physicians – America's second largest physicians group
- Leukemia & Lymphoma Society – America's second largest cancer charity
- American Academy of Family Physicians opposes the use of marijuana except under medical supervision

Other medical organisations recommend a halt to using marijuana as a medicine in U.S.

- The American Society of Addiction Medicine.

History

Ancient China and Taiwan

Cannabis, called *má »* (meaning "hemp; cannabis; numbness") or *dàmá 'Y»* (with "big; great") in Chinese, was used in Taiwan for fibre starting about 10,000 years ago. The botanist Li Hui-Lin wrote that in China, "The use of Cannabis in medicine was probably a very early development. Since ancient men used hemp seed as food, it was quite natural for them to also discover the medicinal properties of the plant." The oldest Chinese pharmacopeia, the (ca. 100 CE) *Shennong Bencaojing* ("Shennong's Materia Medica Classic"), describes *dama* "cannabis".

The flowers when they burst (when the pollen is scattered) are called *mafen* or *mabo*. The best time for gathering is the 7th day of the 7th month. The seeds are gathered in the 9th month. The seeds which have entered the soil are injurious to man. It grows in Taishan (in [Shandong]...). The flowers, the fruit (seed) and the leaves are officinal. The leaves and the fruit are said to be poisonous, but not the flowers and the kernels of the seeds.

In the early 3rd century CE, Hua Tuo was the first person known to use cannabis as an anesthetic. He reduced the plant to powder and mixed it with wine for administration. Elizabeth Wayland Barber says the Chinese evidence "proves a knowledge of the narcotic properties of *Cannabis* at least from the 1st millennium B.C." when *ma* was already used in a secondary meaning of "numbness; senseless." "Such a strong drug, however, suggests that the Chinese pharmacists had now obtained from far to the southwest not THC-bearing *Cannabis sativa* but *Cannabis indica*, so strong it knocks you out cold.

Cannabis is one of the 50 "fundamental" herbs in traditional Chinese medicine, and is prescribed to treat diverse indications.

Every part of the hemp plant is used in medicine; the dried flowers (ÃR), the achenia (a...), the seeds (»·ÁN), the oil (»·¹l), the leaves, the stalk, the root, and the juice. The flowers are recommended in the 120 different forms of (*~feng*) disease, in menstrual disorders, and in wounds. The achenia, which are considered to be poisonous, stimulate the nervous system, and if used in excess, will produce hallucinations and staggering gait. They are prescribed in nervous disorders, especially those marked by local anaesthesia. The seeds, by which is meant the white kernels of the achenia, are used for a great variety of affections, and are considered to be tonic, demulcent, alterative, laxative, emmenagogue, diuretic, anthelmintic, and corrective. They are made into a congee by boiling with water, mixed with wine by a particular process, made into pills, and beaten into a paste. A very common mode of exhibition, however, is by simply eating the kernels. It is said that their continued use renders the flesh firm and prevents old age. They are prescribed internally in fluxes, post-partum difficulties, aconite poisoning, vermillion poisoning, constipation, and obstinate vomiting. Externally they are used for eruptions, ulcers, favus, wounds, and falling of the hair. The oil is used for falling hair, sulfur poisoning, and dryness of the throat. The leaves are considered to be poisonous, and the freshly expressed juice is used as an anthelmintic, in scorpion stings, to stop the hair from falling out and to prevent it from turning grey. They are especially thought to have antiperiodic properties. The stalk, or its bark, is considered to be diuretic, and is used with other drugs in gravel. The juice of the root is used for similar purposes, and is also thought to have a beneficial action in retained placenta and post-partum hemorrhage. An infusion of hemp (for the preparation of which no directions are given) is used as a demulcent drink for quenching thirst and relieving fluxes.

Ancient Egypt

The Ebers Papyrus (ca. 1,550 BCE) from Ancient Egypt describes medical marijuana. Other ancient Egyptian papyri that mention medical marijuana are the Ramesseum III Papyrus (1700 BC), the Berlin Papyrus (1300 BC) and the Chester Beatty Medical Papyrus VI (1300 BC). The ancient Egyptians even used hemp (cannabis) in suppositories for relieving the pain of hemorrhoids. The egyptologist Lise Manniche notes the reference to "plant medical marijuana" in several Egyptian texts, one of which dates back to the eighteenth century BCE.

Ancient India

Surviving texts from ancient India confirm that cannabis' psychoactive properties were recognised, and doctors used it for a variety of illnesses and ailments. These included insomnia, headaches, a whole host of gastrointestinal disorders, and pain: cannabis was frequently used to relieve the pain of childbirth.

Ancient Greece

The Ancient Greeks used cannabis not only for human medicine, but also in veterinary medicine to dress wounds and sores on their horses. In humans, dried leaves of cannabis were used to treat nose bleeds, and cannabis seeds were used to expel tapeworms. The most frequently described use of cannabis in humans was to steep green seeds of cannabis in either water or wine, later taking the seeds out and using the warm extract to treat inflammation and pain resulting from obstruction of the ear.

In the 5th century BCE Herodotus, a Greek historian, described how the Scythians of the Middle East used cannabis in steam baths.

South East Asia

Patani from Asia are primary natural producers of the diuretic, antiemetic, antiepileptic, anti-inflammatory, pain killing and antipyretic properties of *Cannabis sativa*, and used it extensively for 'Kopi Kapuganja' and 'Pecel Ganja', as recreation food, drinks and relaxing medication for centuries.

Medieval Islamic world

In the medieval Islamic world, Arabic physicians made use of the diuretic, antiemetic, antiepileptic, anti-inflammatory, pain killing and antipyretic properties of *Cannabis sativa*, and used it extensively as medication from the 8th to 18th centuries.

Modern History

An Irish physician, William Brooke O'Shaughnessy, is credited with introducing the therapeutic use of cannabis to Western medicine. He was Assistant-Surgeon and Professor of Chemistry at the Medical College of Calcutta, and conducted a cannabis experiment in the 1830s, first testing his preparations on animals, then administering them to patients in order to help treat muscle spasms, stomach cramps or general pain.

Cannabis as a medicine became common throughout much of the Western world by the 19th century, It was used as the primary pain reliever until the invention of aspirin. Modern medical and scientific inquiry began with doctors like O'Shaughnessy and Moreau de Tours, who used it to treat melancholia and migraines, and as a sleeping aid, analgesic and anticonvulsant. At the local level authorities introduced various laws that required the mixtures that contained cannabis, that was not sold on prescription, must be marked with warning labels under the so-called poison laws.

A Swedish lexicon printed in 1912 describes cannabis drug and cannabis extract as a now with us deserted method for medical treatment.

Later in the century, researchers investigating methods of detecting cannabis intoxication discovered that smoking the drug reduced intraocular pressure. In 1955 the antibacterial effects were described at the Palacký University of Olomouc. Since 1971 Lumír Ondøej Hanuš was growing marijuana for his scientific research on two large fields in authority of the University. The marijuana extracts were then used at the University hospital as a cure for aphthae and haze. In 1973 physician Tod H. Mikuriya reignited the debate concerning cannabis as medicine when he published "Marijuana Medical Papers". High intraocular pressure causes blindness in glaucoma patients, so he hypothesized that using the drug could prevent blindness in patients. Many Vietnam Warveterans also found that the drug prevented muscle spasms caused by spinal injuries suffered in battle. Later medical use focused primarily on its role in preventing the wasting syndromes and chronic loss of appetite associated with chemotherapy and AIDS, along with a variety of rare muscular and skeletal disorders.

In 1964, Dr. Albert Lockhart and Manley West began studying the health effects of traditional marijuana use in Jamaican communities. They discovered that Rastafarians had unusually low glaucoma rates and local fishermen were washing their eyes with cannibis extract in the belief that it would improve their sight. Lockhart and West developed, and in 1987 gained permission to market, the pharmaceutical Canasol: one of the first to cannabis extracts. They continued to work with cannabis throughout the years, developing more pharmaceuticals and eventually receiving the Jamaican Order of Merit for their work.

Later, in the 1970s, a synthetic version of THC was produced and approved for use in the United States as the drug Marinol. It was delivered as a capsule, to be swallowed. Patients complained that the violent nausea associated with chemotherapy made swallowing capsules difficult. Further, along with ingested cannabis, capsules are harder to dose-titrate accurately than smoked cannabis because their onset of action is so much slower. Smoking has remained the route of choice for many patients because its onset of action provides almost immediate relief from symptoms and because that fast onset greatly simplifies titration. For these reasons, and because of the difficulties arising from the way cannabinoids are metabolised after being ingested, oral dosing is probably the least satisfactory route for cannabis administration. Relatedly, some studies have indicated that at least some of the beneficial effects that cannabis can provide may derive from synergy among the multiplicity of cannabinoids and other chemicals present in the dried plant material. Such synergy is, by definition, impossible with respect to the use of single-cannabinoid drugs like Marinol.

During the 1970s and 1980s, six U.S. states' health departments performed studies on the use of medical cannabis. These are widely considered some of the most useful and pioneering studies on the subject. Voters in eight states showed their support for cannabis prescriptions or recommendations given by physicians between 1996 and 1999, including Alaska, Arizona, California, Colorado, Maine, Michigan, Nevada, Oregon, and Washington, going against policies of the federal government.

In May 2001, "The Chronic Cannabis Use in the Compassionate Investigational New Drug Program: An Examination of Benefits and Adverse Effects of Legal Clinical Cannabis" (Russo, Mathre, Byrne et al.) was completed. This three-day examination of major body functions of four of the five living US federal cannabis patients found "mild pulmonary changes" in two patients.

On October 7, 2003, a U.S. patent US 6630507 entitled "Cannabinoids as Antioxidants and Neuroprotectants" was awarded to the United States Department of Health and Human Services, based on research done at the National Institute of Mental Health (NIMH), and the National Institute of Neurological Disorders and Stroke (NINDS). This patent claims that cannabinoids are "useful in the treatment and prophylaxis of wide variety of oxidation associated diseases, such as ischemic, age-related, inflammatory and autoimmune diseases. The cannabinoids are found to have particular application as neuroprotectants, for example in limiting neurological damage following ischemic insults, such as stroke and trauma, or in the treatment of neurodegenerative diseases, such as Alzheimer's disease, Parkinson's disease and HIV dementia."

National and International Regulations

Cannabis is in Schedule IV of the United Nations´ Single Convention on Narcotic Drugs, making it subject to special restrictions. Article 2 provides for the following, in reference to Schedule IV drugs:

A Party shall, if in its opinion the prevailing conditions in its country render it the most appropriate means of protecting the public health and welfare, prohibit the production, manufacture, export and import of, trade in, possession or use of any such drug except for amounts which may be necessary for medical and scientific research only, including clinical trials therewith to be conducted under or subject to the direct supervision and control of the Party. The convention thus allows countries to outlaw cannabis for all non-research purposes but lets nations choose to allow medical and scientific purposes if they believe total prohibition is not the most appropriate means of protecting health and welfare. The convention requires that states that permit the production or use of medical cannabis must operate a licensing system for all cultivators, manufacturers and distributors and ensure that the total cannabis market of the state shall not exceed that required "for medical and scientific purposes."

Austria

In Austria both Ä⁹-THC and pharmaceutical preparations containing Ä⁹-THC are listed in annex V of the Narcotics Decree (*Suchtgiftverordnung*). Compendial formulations are manufactured upon prescription according to the German *Neues Rezeptur-Formularium*. On July 9, 2008, the Austrian Parliament approved cannabis cultivation for scientific and medical uses. Cannabis cultivation is controlled by the Austrian Agency for Health and Food Safety (*Österreichische Agentur für Gesundheit und Ernährungssicherheit, AGES*).

Canada

In Canada, the regulation on access to marijuana for medical purposes, established by Health Canada in July 2001, defines two categories of patients eligible for access to medical cannabis. Category 1 covers any symptoms treated within the context of providing compassionate end-of-life care or the symptoms associated with medical conditions listed below:

- severe pain and/or persistent muscle spasms from multiple sclerosis, from a spinal cord injury, from spinal cord disease,
- severe pain, cachexia, anorexia, weight loss, and/or severe nausea from cancer or HIV/AIDS infection,
- severe pain from severe forms of arthritis, or
- seizures from epilepsy.

Category 2 is for applicants who have debilitating symptom(s) of medical condition(s), other than those described in Category 1. The application of eligible patients must be supported by a medical practitioner. The cannabis distributed by Health Canada is provided under the brand CannaMed by the company Prairie Plant Systems Inc. In 2006, 420 kg of CannaMed cannabis was sold, representing an increase of 80% over the previous year. However, patients complain of the single strain selection as well as low potency, providing a pre-ground product put through a wood chipper (which deteriorates rapidly) as well as gamma irradation and foul taste and smell.

It is also legal for patients approved by Health Canada to grow their own cannabis for personal consumption, and it's possible to obtain a production license as a person designated by a patient. Designated producers were permitted to grow a cannabis supply for only a single patient, however. That regulation and related restrictions on supply were found unconstitutional by the Federal Court of Canada in January, 2008. The court found that these regulations did not allow a sufficient legal supply of medical cannabis, and thus forced many patients to purchase their medicine from unauthorised, black market sources. This was the eighth time in the previous ten years that the courts ruled against Health Canada's regulations restricting the supply of the medicine.

In May, 2009, Health Canada revised their earlier regulations to permit licensed, designated producers to grow cannabis for a maximum of two patients. The move was called a "mockery" of the court's intention by lawyer Ron Marzel, who represented plaintiffs in the successful challenge in Federal Court to Health Canada's previously existing rules. Marzel has announced plans to ask the court to overturn all prohibitions on cannabis use if Health Canada refuses to create regulations that will allow an adequate legal supply for use by medically-authorised patients.

In Canada there are four forms of medical marijuana. The first one is a cannabis extract called Sativex that contains THC and cannabidiol in a spray form. The second is a synthetic or manmade THC called dronabinol marketed as Marinol. The third also a synthetic version of THC called nabilone that is called Cesamet on the markets. The fourth product is the herbal form of cannabis often referred to as marijuana.

Spain

In Spain, since the late 1990s and early 2000s, medical cannabis underwent a process of progressive decriminalisation and legalisation. The parliament of the region of Catalonia is the first in Spain have voted unanimously in 2001 legalising medical marijuana, it is quickly followed by parliaments of Aragon and the Balearic Islands. The Spanish Penal Code prohibits the sale of cannabis but it does not prohibit consumption (although consumption on the street is fined). Until early 2000, the Penal Code did not distinguish between therapeutic use of cannabis and recreational use, however, several court decisions show that this distinction is increasingly taken into account by the judges. From 2006, the sale of seed is legalised, the sale and public consumption remains illegal, and private cultivation and use are permitted.

Several studies have been conducted to study the effects of cannabis on patients suffering from diseases like cancer, AIDS, multiple sclerosis, seizures or asthma. This research was conducted by various Spanish agencies at the Universidad Complutense de Madrid headed by Manuel Guzman, the hospital of La Laguna in Tenerife led neurosurgeon Luis González Feria or the University of Barcelona.

Several cannabis consumption clubs and user associations have been established throughout Spain. These clubs, the first of which was created in 1991, are non-profit associations who grow cannabis and sell it at cost to its members. The legal status of these clubs is uncertain: in 1997, four members of the first club, the Barcelona Ramón Santos Association of Cannabis Studies, were sentenced to 4 months in prison and a 3000 euro fine, while at about the same time, the court of Bilbao ruled that another club was not in violation of the law. The Andalusian regional government also commissioned a study by criminal law professors on the "Therapeutic use of cannabis and the creation of establishments of acquisition and consumption. The study concluded that such clubs are legal as long as they distribute only to a restricted list of legal adults, provide only the amount of drugs necessary for immediate consumption, and not earn a profit. The Andalusian government never formally accepted these guidelines and the legal situation of the clubs remains insecure. In 2006 and 2007, members of these clubs were acquitted in trial for possession and sale of cannabis and the police were ordered to return seized crops.

United Kingdom

In the United Kingdom, if you are arrested or taken to court for possession of cannabis, you are asked if there are any mitigating factors to explain why it is in your possession. It is unknown whether this is solely a formality, or if an excuse of medical usage has ever been used successfully to reduce the penalty issued. However, in the United Kingdom, possession of small quantities of cannabis does not usually warrant an arrest or court appearance (street cautions or fines are often given out instead). Under UK law, certain cannabinoids are permitted medically, but these are strictly controlled with many provisos under the Misuse of drugs act 1971 (in the 1985 amendments). The British Medical Associations official stance is "users of cannabis for medical purposes should be aware of the risks, should enroll for clinical trials, and should talk to their doctors about new alternative treatments; but we do not advise them to stop."

United States

In the United States federal level of government, cannabis *per se* has been made criminal by implementation of the Controlled Substances Act which classifies marijuana as a Schedule I drug, the strictest classification on par with heroin, LSD and Ecstasy, and the Supreme Court ruled in 2005 that the Commerce Clause of the U.S. Constitution allowed the government to ban the use of cannabis, including medical use. The United States Food and Drug Administration states "marijuana has a high potential for abuse, has no currently accepted medical use in treatment in the United States, and has a lack of accepted safety for use under medical supervision".

Sixteen states have legalised medical marijuana: Alaska, Arizona, California, Colorado, Hawaii, Maine, Michigan, Montana, Nevada, New Jersey, New Mexico, Oregon, Rhode Island, Vermont, Virginia, Washington; and Washington D.C. Maryland allows for reduced or no penalties if cannabis use has a medical basis. Despite its legality in Washington, an employee can be still be fired if they test positive on a drug test, despite having a doctor's recommendation. California, Colorado, New Mexico, Maine, Rhode Island, Montana, and Michigan are currently the only states to utilise dispensaries to sell medical cannabis. California's medical marijuana industry took in about $2 billion a year and generated $100 million in state sales taxes during 2008 with an estimated 2,100 dispensaries, co-operatives, wellness clinics and taxi delivery services in the sector colloquially known as "cannabusiness".

On 19 October 2009 the US Deputy Attorney General issued a US Department of Justice memorandum to "All United States Attorneys" providing clarification and guidance to federal prosecutors in US States that have enacted laws authorising the medical use of marijuana. The document is intended solely as "a guide to the exercise of investigative and prosecutorial discretion and as guidance on resource allocation and federal priorities." The US Deputy Attorney General David W. Ogden provided seven criteria, the application of which acts as a guideline to prosecutors and federal agents to ascertain whether a patients use, or their caregivers provision, of medical marijuana "represents part of a recommended treatment regiment consistent with applicable state law", and recommends against prosecuting patients using medical cannabis products according to state laws. Not applying those criteria, the Dep. Attorney General Ogden concludes, would likely be "an inefficient use of limited federal resources". The memorandum does not change any laws. Sale of cannabis remains illegal under federal law. The U.S. Food and Drug Administration's position, that marijuana has no accepted value in the treatment of any disease in the United States, has also remained the same.

The Health and Human Services Division of the federal government holds the patent US 6630507 for medical marijuana. The patent, "Cannabinoids as antioxidants and neuroprotectants", issued October 2003 reads: "Cannabinoids have been found to have antioxidant properties, unrelated to NMDA receptor antagonism. This new found property makes cannabinoids useful in the treatment and prophylaxis of wide variety of oxidation associated diseases, such as ischemic, age-related, inflammatory and autoimmune diseases. The cannabinoids are found to have particular application as neuroprotectants, for example in limiting neurological damage following ischemic insults, such as stroke and trauma, or in the treatment of neurodegenerative diseases, such as Alzheimer's disease, Parkinson's disease and HIV dementia..."

Chapter 2

History of Plant Systematics

The history of plant systematics—the biological classification of plants—stretches from the work of ancient Greek to modern evolutionary biologists. As a field of science, plant systematics came into being only slowly, early plant lore usually being treated as part of the study of medicine. Later, classification and description was driven by natural history and natural theology. Until the advent of the theory of evolution, nearly all classification was based on the scala naturae. The professionalisation of botany in the 18th and 19th century marked a shift toward more holistic classification methods, eventually based on evolutionary relationships.

Antiquity

Historians of botany trace the origins of botanical classification with folk taxonomy and ancient Greece.

Theophrastus's (372–287 BC), a student of Aristotle, produced *Historia Plantarum*, the earliest surviving treatise on plants, where he listed the names of over 500 plant species. However he did not articulate a formal classification scheme; instead he relied on the common groupings of folklore combined with growth form: tree shrub; undershrub; or herb.

The *Materia medica* of Dioscorides was also an important early compendium of plant descriptions (over five hundred); it was in use from its publication in the 1st century until the 16th century.

Early Modern Period

In the 16th century, works by Otto Brunfels, Hieronymus Bock, and Leonhart Fuchs helped to revive interest in natural history based on first-hand observation; Bock in particular included environmental and life cycle information in his descriptions. With the influx of exotic species in the Age of Exploration, the number of known species expanded rapidly, but most authors were far more interested in the medicinal properties of individual plants than an overarching classification system. Later influential Renaissance books include those of Caspar Bauhin and Andrea Cesalpino. Bauhin described over 6000 plants, which he arranged into 12 books and 72 sections based on a wide range of common characteristics. Cesalpino based his system on the structure of the organs of fructification, using the Aristotelian technique of logical division.

In the late 17th century, the most influential classification schemes were those of English botanist and natural theologian John Ray and French botanist Joseph Pitton de Tournefort. Ray, who listed over 18,000 plant species in his works, is credited with establishing the monocot/dicot division and some of his groups — mustards, mints, legumes and grasses — stand today (though under modern family names). Tournefort used an artificial system based on logical division which was widely adopted in France and elsewhere in Europe up until Linnaeus.

The book that had an enormous accelerating effect on the science of plant systematics was *Species Plantarum* (1753) by Linnaeus. It presented a complete list of the plant species then known to Europe, ordered for the purpose of easy identification using the number and arrangement of the male and female sexual organs of the plants. Of the groups in this book, the highest rank that continues to be used today is the genus. The consistent use of binomial nomenclature along with a complete listing of all plants provided a huge stimulus for the field.

Although meticulous, the classification of Linnaeus served merely as an identification manual; it was based on phenetics and did not regard evolutionary relationships among species. It assumed that plant species were given by God and that what remained for humans was to recognise them and use them (a Christian reformulation of the *scala naturae* or *Great Chain of Being*). Linnaeus was quite aware that the arrangement of species in the *Species Plantarum* was not a natural system, i.e. did not express relationships. However he did present some ideas of plant relationships elsewhere.

Modern and Contemporary Periods

Significant contributions to plant classification came from de Jussieu (inspired by the work of Adanson) in 1789 and the early nineteenth century saw the start of work by de Candolle, culminating in the *Prodromus*.

A major influence on plant systematics was the theory of evolution (Charles Darwin published *Origin of Species* in 1859), resulting in the aim to group plants by their phylogenetic relationships. To this was added the interest in plant anatomy, aided by the use of the light microscope and the rise of chemistry, allowing the analysis of secondary metabolites.

Currently, the strict use of epithets in botany, although regulated by international codes, is considered unpractical and outdated. The very notion of species, the fundamental classification unit, is often up to subjective intuition and thus can not be well defined. As a result, estimate of the total number of existing "species" (ranging from 2 million to 100 million) becomes a matter of preference.

While scientists have agreed for some time that a functional and objective classification system must reflect actual evolutionary processes and genetic relationships, the technological means for creating such a system did not exist until recently. In the 1990s DNA technology saw immense progress, resulting in unprecedented accumulation of DNA sequence data from various genes present in compartments of plant cells. In 1998 a ground-breaking classification of the angiosperms (the APG system) consolidated molecular phylogenetics (and especially cladistics or phylogenetic systematics) as the best available method. For the first time relatedness could be measured in real terms, namely similarity of the molecules comprising the genetic code.

Timeline of Publications

*   Theophrastus. *Historia Plantarum.*
*   A. Cesalpino (1583). *De plantis libri XVI.*
*   John Ray (1686). *Historia Plantarum.*
*   Linnaeus (1753). *Species Plantarum.*
*   M. Adanson (two volumes, 1763). *Familles des plantes.*
*   A.L. de Jussieu (1789). *Genera Plantarum, secundum ordines naturales disposita juxta methodum in Horto Regio Parisiensi exaratam.*
*   A. P. de Candolle et al. (1824–1873). *Prodromus systemati naturalis regni vegetabilis sive enumeratio contracta ordinum, generum specierumque plantarum huc usque cognitarum, juxta methodi naturalis normas digesta.*
*   Lindley, John (1846). *The Vegetable Kingdom.*

Amaranth

Amaranthus, collectively known as amaranth, is a cosmopolitan genus of herbs. Approximately 60 species are recognised, with inflorescences and foliage ranging from purple and red to gold. Members of this genus share many characteristics and uses with members of the closely related genus *Celosia*.

Although several species are often considered weeds, people around the world value amaranths as leaf vegetables, cereals, and ornamentals.

The ultimate root of "amaranth" is the Greek *amarantos* "unfading" with the Greek word for "flower" *anthos* factoring into the word's development as "amaranth"- the more correct "amarant" is an archaic variant.

Systematics

Amaranthus shows a wide variety of morphological diversity among and even within certain species. Although the family (Amaranthaceae) is distinctive, the genus has few distinguishing characters among the 70 species included. This complicates taxonomy and *Amaranthus* has generally been considered among systematists as a "difficult" genus.

Formerly, Sauer (1955) classified the genus into 2 sub-genera, differentiating only between monoecious and dioecious species: *Acnida* (L.) Aellen ex K.R. Robertson and *Amaranthus*. Although this classification was widely accepted, further infrageneric classification was (and still is) needed to differentiate this widely diverse group.

Currently, *Amaranthus* includes 3 recognised sub-genera and 70 species, although species numbers are questionable due to hybridisation and species concepts. Infrageneric classification focuses on inflorescence, flower characters and whether a species is monoecious/dioecious, as in the Sauer (1955) suggested classification. A modified infrageneric classification of *Amaranthus* was published by Mosyakin & Robertson (1996) and includes 3 subgenera: *Acnida*, *Amaranthus*, and *Albersia*. The taxonomy is further differentiated by sections within each of the sub-genera.

Species:

- *Amaranthus acanthochiton* – greenstripe
- *Amaranthus acutilobus* – sharp-lobe amaranth; is a synonym of *Amaranthus viridis*
- *Amaranthus albus* – white pigweed, prostrate pigweed, pigweed amaranth
- *Amaranthus arenicola* – sandhill amaranth
- *Amaranthus australis* – southern amaranth
- *Amaranthus bigelovii* – Bigelow's amaranth

- *Amaranthus blitoides* – mat amaranth, prostrate amaranth, prostrate pigweed
- *Amaranthus blitum* – purple amaranth
- *Amaranthus brownii* – Brown's amaranth
- *Amaranthus californicus* – California amaranth, California pigweed
- *Amaranthus cannabinus* – tidal-marsh amaranth
- *Amaranthus caudatus* – love-lies-bleeding, pendant amaranth, tassel flower, quilete
- *Amaranthus chihuahuensis* – chihuahuan amaranth
- *Amaranthus chlorostachys*
- *Amaranthus crassipes* – spreading amaranth
- *Amaranthus crispus* – crispleaf amaranth
- *Amaranthus cruentus* – purple amaranth, red amaranth, Mexican grain amaranth
- *Amaranthus deflexus* – large-fruit amaranth
- *Amaranthus dubius* – spleen amaranth, khada sag
- *Amaranthus fimbriatus* – fringed amaranth, fringed pigweed
- *Amaranthus floridanus* – Florida amaranth
- *Amaranthus gangeticus* – elephant head amaranth
- *Amaranthus graecizans*
- *Amaranthus greggii* – Gregg's amaranth
- *Amaranthus hybridus* – smooth amaranth, smooth pigweed, red amaranth
- *Amaranthus hypochondriacus* – Prince-of-Wales-feather, princess feather
- *Amaranthus leucocarpus*
- *Amaranthus lineatus* – Australian amaranth
- *Amaranthus lividus*
- *Amaranthus mantegazzianus* – Quinoa de Castilla
- *Amaranthus minimus*
- *Amaranthus muricatus* – African amaranth
- *Amaranthus obcordatus* – Trans-Pecos amaranth
- *Amaranthus oleraceous* – Kosala Sag
- *Amaranthus palmeri* – Palmer's amaranth, palmer pigweed, careless weed

- *Amaranthus paniculus* – Reuzen amarant
- *Amaranthus polygonoides* – tropical amaranth
- *Amaranthus powellii* – green amaranth, Powell amaranth, Powell pigweed
- *Amaranthus pringlei* – Pringle's amaranth
- *Amaranthus pumilus* – seaside amaranth
- *Amaranthus quitensis* – ataco, sangorache
- *Amaranthus retroflexus* – red-root amaranth, redroot pigweed, common amaranth
- *Amaranthus rudis* – tall amaranth, common waterhemp
- *Amaranthus scleropoides* – bone-bract amaranth
- *Amaranthus spinosus* – spiny amaranth, prickly amaranth, thorny amaranth
- *Amaranthus standleyanus*
- *Amaranthus thunbergii* – Thunberg's amaranth
- *Amaranthus torreyi* – Torrey's amaranth
- *Amaranthus tricolour* – Joseph's-coat
- *Amaranthus tuberculatus* – rough-fruit amaranth, tall waterhemp
- *Amaranthus viridis* – slender amaranth, green amaranth
- *Amaranthus watsonii* – Watson's amaranth
- *Amaranthus wrightii* – Wright's amaranth.

Uses

Amaranth Seed

Several species are raised for amaranth "grain" in Asia and the Americas. This should more correctly be termed "pseudograin". Amaranth grain contains no gluten and is safe to consume for individuals with coeliac disease.

Ancient amaranth grains still used to this day include the three species, *Amaranthus caudatus*, *Amaranthus cruentus*, and *Amaranthus hypochondriacus*. Although amaranth was cultivated on a large scale in ancient Mexico, Guatemala, and Peru, nowadays it is only cultivated on a small scale there, along with India, China, Nepal, and other tropical countries; thus, there is potential for further cultivation in those countries, as well as in the U.S.

In a 1977 article in *Science*, amaranth was described as "the crop of the future." It has been proposed as an inexpensive native crop that could be cultivated by indigenous people in rural areas for several reasons:

1. It is easily harvested.
2. It is highly tolerant of arid environments, which are typical of most subtropical and some tropical regions, and
3. Its seeds are a good source of protein, rich in essential amino acids such as lysine, while being a poor source of essential amino acids such as leucine and threonine. Common grains such as wheat and corn are rich in amino acids that amaranth lacks; thus, amaranth and other grains can complement each other.
4. The seeds of *Amaranthus* species contain about thirty percent more protein than cereals like rice, sorghum and rye. In cooked and edible forms, amaranth is competitive with wheat germ and oats- higher in some nutrients, lower in others.
5. It is easy to cook. As befits its weedy life history, amaranth grains grow very rapidly and their large seedheads can weigh up to 1 kilogram and contain a half-million seeds in three species of amaranth.

Kiwicha, as amaranth is known today in the Andes, was one of the staple foodstuffs of the Incas. Known to the Aztecs as huautli, it is thought to have represented up to 80% of their caloric consumption before the conquest. Another important use of amaranth throughout Mesoamerica was to prepare ritual drinks and foods. To this day, amaranth grains are toasted much like popcorn and mixed with honey, molasses or chocolate to make a treat called *alegría*, meaning "joy" in Spanish. Diego Duran described the festivities for Huitzilopochtli, a blue hummingbird god. (Real hummingbirds feed on amaranth flowers.) The Aztec month of Panquetzaliztli (7 December to 26 December) was dedicated to Huitzilopochtli. People decorated their homes and trees with paper flags; there were ritual races, processions, dances, songs, prayers, and finally human sacrifices. This was one of the more important Aztec festivals, and the people prepared for the whole month. They fasted or ate very little; a statue of the god was made out of amaranth (*huautli*) seeds and honey, and at the end of the month, it was cut into small pieces so everybody could eat a little piece of the god. After the Spanish conquest, cultivation of amaranth was outlawed, while some of the festivities were subsumed into the Christmas celebration.

Because of its importance as a symbol of indigenous culture, and because it is very palatable, easy to cook, and its protein particularly well suited to human nutritional needs, interest in grain amaranth (especially *A. cruentus*and *A. hypochondriacus*) revived in the 1970s. It was recovered in Mexico from wild varieties and is now commercially cultivated. It is a popular snack sold in Mexico City and other parts of Mexico, sometimes mixed with chocolate or puffed rice, and its use has spread to Europe and parts of North America. Amaranth and quinoa are called pseudograins because of their flavour and cooking similarities to grains. These are dicot plant seeds, and both contain exceptionally complete protein for plant sources. Besides protein, amaranth grain provides a good source of dietary fibre and dietary minerals such as iron, magnesium, phosphorus, copper, and especially manganese.

Scientific studies suggest Amaranth grain is a good source of essential amino acid lysine, something other grains are low in. Amaranth is not a complete source of essential amino acids. For example, amaranth is limiting in leucine and threonine- essential amino acids that are abundant in other grains. Amaranth may therefore be a promising supplement to other grains.

Amaranth seed flour has been evaluated as an additive to wheat flour by food specialists. To determine palatability, different levels of amaranth grain flour were mixed with the wheat flour and baking ingredients (1% salt, 2.5% fat, 1.5% yeast, 10% sugar and 52–74% water), fermented, molded, pan-proved and baked. The baked products were evaluated for loaf volume, moisture content, colour, odor, taste and texture. The amaranth containing products were then compared with bread made from 100% wheat flour. The loaf volume decreased by 40% and the moisture content increased from 22 to 42% with increase in amaranth grain flour. The study found that the sensory scores of the taste, odor colour and texture decreased with increasing amounts of amaranth. Generally, above 15% amaranth grain flour, there were significant differences in the evaluated sensory qualities and the high amaranth-containing product was found to be of unacceptable palatability to the population sample that evaluated the baked products.

Amaranth grain is a crop of moderate importance in the Himalayas.

Leaves, Roots, and Stems

Amaranth species are cultivated and consumed as a leaf vegetable in many parts of the world. There are four species of *Amaranthus* documented as cultivated vegetables in eastern Asia: *Amaranthus cruentus*, *Amaranthus blitum*, *Amaranthus dubius*, and *Amaranthus tricolour*.

In Indonesia and Malaysia, leaf amaranth is called *bayam*, while the Tagalogs in the Philippines call the plant *alocon*. In the state of Uttar Pradesh and Bihar in India, it is called Chaulai and is a popular green leafy vegetable (referred to in the class of vegetable preparations called Saag). It is called Chua in Kumaun area of Uttarakhand, where it is a popular red-green vegetable. In Karnataka state in India it is used to prepare curries like Hulee, palya, Majjigay-hulee and so on. In the state of Kerala, it's called 'Cheera' and is consumed by stir-frying the leaves with spices and red chillies to make 'Cheera Thoran'. In Tamilnadu State, it is called ®Á³È·Í·À°È and is regularly consumed as a favourite dish, where the greens are steamed, and mashed, with light seasoning of salt, red chillis and cumin. It is called *keerai masial*. In Andhra Pradesh, India, this leaf is added in preparation of a popular dal called *thotakura pappu* $J B0 **M*A (Telugu). In Maharashtra, it is called as "Shravani Maath" and it is available in both red and white colour. In Orissa, it is called as "Khada saga", it is used to prepare 'Saga Bhaja', in which the leaf is fried with chillies and onions.

The root of mature amaranth is an excellent vegetable. It is white and cooked with tomatoes or tamarind gravy. It has a milky taste and is alkaline.

In China, the leaves and stems are used as a stir-fry vegetable, or in soups, and called *yin choi*. Amaranth greens are believed to help enhance eyesight. In Vietnam, it is called *rau dÁn* and is used to make soup. There are two species popular as edible vegetable in Vietnam: *dÁn ðÏ*- amaranthus tricolour and *dÁn cõm* or *dÁn tr¯ng*- amaranthus viridis.

A traditional food plant in Africa, amaranth has the potential to improve nutrition, boost food security, foster rural development and support sustainable landcare. In East Africa, Amaranth leaf is known in chewa as *bonongwe,* and in Swahilias *mchicha,* as *terere* in Kikuyu, Meru and Embu; and as *telele* in Kamba. In Bantu regions of Uganda it is known as *doodo*. It is recommended by some doctors for people having low red blood cell count. It is also known among the Kalenjin as a drought crop (*chepkerta*). In Lingala (spoken in the Democratic Republic of the Congo and the Republic of Congo), it is known as *l[ngal[nga* or *bít[kut[ku*. In Nigeria, it is a common vegetable and goes with all Nigerian starch dishes. It is known in Yoruba as *efo tete* or *arowo jeja* (meaning "we have money left over for fish"). In the Caribbean, the leaves are called *bhaji in Trinidad or callaloo in Jamaica* and stewed with onions, garlic and tomatoes, or sometimes used in a soup called pepperpot soup.

In Greece, green amaranth (*Amaranthus viridis*) is a popular dish and is called *vlita* or *vleeta*. It's boiled, then served with olive oil and lemon like a salad, usually alongside fried fish. Greeks stop harvesting the plant (which usually grows wild) when it starts to bloom at the end of August.

In Sri Lanka, it is called "koora thampala". Sri Lankans cook it and eat it with rice. Fiji Indians call it choraiya bhaji.

Dyes

The flowers of the 'Hopi Red Dye' amaranth were used by the Hopi (a tribe in the western United States) as the source of a deep red dye. There is also a synthetic dye that has been named "amaranth" for its similarity in colour to the natural amaranth pigments known as betalains. This synthetic dye is also known as Red No. 2 in North America and E123 in the European Union.

Ornamentals

The genus also contains several well-known ornamental plants, such as *Amaranthus caudatus* (love-lies-bleeding), a native of India and a vigorous, hardy annual with dark purplish flowers crowded in handsome drooping spikes. Another Indian annual, *A. hypochondriacus* (prince's feather), has deeply veined lance-shaped leaves, purple on the under face, and deep crimson flowers densely packed on erect spikes.

Amaranths are recorded as food plants for some Lepidoptera (butterfly and moth) species including the nutmeg moth and various case-bearer moths of the genus *Coleophora*: *C. amaranthella*, *C. enchorda* (feeds exclusively on *Amaranthus*), *C. immortalis* (feeds exclusively on *Amaranthus*), *C. lineapulvella* and *C. versurella* (recorded on *A. spinosus*).

Nutritional Value

Amaranth greens, also called Chau lai (Hindi) and Chu or Chua (Kumauni), Chinese spinach, hinn choy or yin tsoi (simplified Chinese: Ë,Üƒ; traditional Chinese: pinyin: callaloo, dhantinasoppu (Kannada); $KB0(Telugu); Rajgira (Marathi); (Tamil); cheera (Malayalam); bayam (Indonesian); phak khom (Thai); tampala, or quelite (Oriya); Khada Saga, are a common leaf vegetable throughout the tropics and in many warm temperate regions.

Cooked amaranth leaves are a good source of vitamin A, vitamin C, and folate; they are also a complementing source of other vitamins such as thiamine, niacin, and riboflavin, plus some dietary minerals including calcium, iron, potassium,zinc, copper, and manganese. Cooked amaranth grains are a complementing source of thiamine, niacin, riboflavin, folate, and dietary minerals including calcium, iron, magnesium, phosphorus, zinc, copper, and manganese- comparable to common grains such as wheat germ, oats and others.

Amaranth seeds contain lysine, an essential amino acid, limiting in other grains or plant sources. Most fruits and vegetables do not contain a complete set of amino acids, and thus different sources of protein must be used. Amaranth too is limiting in some essential amino acids, such as leucine and threonine. Amaranth seeds are therefore promising complement to common grains such as wheat germ, oats, corn because these common grains are abundant sources of essential amino acids found to be limited in amaranth.

Amaranth may be a promising source of protein to those who are gluten sensitive, because unlike the protein found in grains such as wheat and rye, its protein does not contain gluten. According to a 2007 report, amaranth compares well in nutrient content with gluten-free vegetarian options such as buckwheat, corn, millet, wild rice, oats and quinoa.

Several studies have shown that like oats, amaranth seed or oil may be of benefit for those with hypertension and cardiovascular disease; regular consumption reduces blood pressure and cholesterol levels, while improving antioxidant status and some immune parameters. While the active ingredient in oats appears to be water-soluble fibre, amaranth appears to lower cholesterol via its content of plant stanols and squalene.

Amaranth remains an active area of scientific research for both human nutritional needs and foraging applications. Over 100 scientific studies suggest a somewhat conflicting picture on possible anti-nutritional and toxic factors in amaranth, more so in some particular strains of amaranth. Lehmann, in a review article, identifies some of these reported anti-nutritional factors in amaranth to be phenolics, saponins, tannins, phytic acid, oxalates, protease inhibitors, nitrates, polyphenols and phytohemagglutinins. Of these, oxalates and nitrates are of more concern when amaranth grain is used in foraging applications. Some studies suggest thermal processing of amaranth, particularly in moist environment, prior to its preparation in food and human consumption may be a promising way to reduce the adverse effects of amaranth's anti-nutritional and toxic factors.

A one-to-one comparison of cooked amaranth with cooked wild rice and with whole grain wheat flour suggests:

- the nutrition content of cooked amaranth is higher in some, lower in other essential nutrients in comparison to wild rice.
- the nutrition content of cooked amaranth is higher in few, lower in most other essential nutrients in comparison to whole grain wheat.

As a Weed

Not all amaranth plants are cultivated. Most of the species from *Amaranthus* are summer annual weeds and are commonly referred to as pigweeds. These species have an extended period of germination, rapid growth, and high rates of seed production, and have been causing problems for farmers since the mid-1990s. This is partially due to the reduction in tillage, reduction in herbicidal use and the evolution of herbicidal resistance in several species where herbicides have been applied more often. The following 9 species of *Amaranthus* are considered invasive and noxious weeds in the U.S and Canada: *A. albus*, *A. blitoides*, *A. hybridus*, *A. palmeri*, *A. powellii*, *A. retroflexus*, *A. spinosus*, *A. tuberculatus*, and *A. viridis*.

A new herbicide-resistant strain of Amaranthus palmeri or Palmer amaranth has appeared; it is Glyphosate-resistant and so cannot be killed by the widely used Roundup herbicide. Also, this plant can survive in tough conditions. This could be of particular concern to cotton farmers using Roundup Ready cotton. The species *Amaranthus palmeri* (Palmer amaranth) causes the greatest reduction in soybean yields and has the potential to reduce yields by 17-68% in field experiments. Palmer amaranth is among the "top five most troublesome weeds" in the southeast and has already evolved resistances to dinitroanilines and acetolactate synthase inhibitors. This makes the proper identification of *Amaranthus* species at the seedling stage essential for agriculturalists. Proper weed control needs to be applied before the species successfully colonises in the crop field and causes significant yield reductions.

Beneficial Weed

Pigweed can be a beneficial weed, as well as a companion plant, serving as a trap for leaf miners and some other pests, as well as sheltering ground beetles (which prey upon insect pests) and breaking up hard soil for more delicate neighbouring plants.

Myth, Legend and Poetry

The word amaranth comes from the Greek word *amarantos*, meaning "unwithering". The word was applied to amaranth because it did not soon fade and so symbolised immortality. "Amarant" is a more correct, albeit archaic form, chiefly used in poetry. The current spelling, *amaranth*, seems to have come from folk etymology that assumed the final syllable derived from the Greek word *anthos* ("flower"), common in botanical names.

Aesop's Fables (6th century BC) compares the rose to the amaranth to illustrate the difference in fleeting and everlasting beauty:

A Rose and an Amaranth blossomed side by side in a garden, and the Amaranth said to her neighbour,

> *"How I envy you your beauty and your sweet scent!*
> *No wonder you are such a universal favourite."*
> *But the Rose replied with a shade of sadness in her voice,*
> *"Ah, my dear friend, I bloom but for a time:*
> *my petals soon wither and fall, and then I die.*
> *But your flowers never fade, even if they are cut;*
> *for they are everlasting."*

Or in story form:

An amaranth planted in a garden near a Rose-Tree, thus addressed it: "What a lovely flower is the Rose, a favourite alike with Gods and with men. I envy you your beauty and your perfume." The Rose replied, "I indeed, dear Amaranth, flourish but for a brief season! If no cruel hand pluck me from my stem, yet I must perish by an early doom. But thou art immortal and dost never fade, but bloomest for ever in renewed youth."

Thus, in John Milton's epic poem *Paradise Lost* (1667), iii. 353:

> *"Immortal amarant, a flower which once*
> *In paradise, fast by the tree of life,*
> *Began to bloom; but soon for man's offence*
> *To heaven removed, where first it grew, there grows,*
> *And flowers aloft, shading the fount of life,*
> *And where the river of bliss through midst of heaven*
> *Rolls o'er elysian flowers her amber stream:*
> *With these that never fade the spirits elect*
> *Bind their resplendent locks."*

Samuel Taylor Coleridge, in *Work Without Hope* (1825), also refers to the herb, likely referencing Milton's earlier work. (ll 7-10 excerpted):

Yet well I ken the banks where Amaranths blow,

Have traced the fount whence streams of nectar flow.

Bloom, O ye Amaranths! bloom for whom ye may,

For me ye bloom not! Glide, rich streams, away!

In his dialogue 'Aesop and Rhodopè', published in 1844, Walter Savage Landor wrote:

There are no fields of amaranth on this side of the grave: there are no voices, O Rhodopè, that are not soon mute, however tuneful: there is no name, with whatever emphasis of passionate love repeated, of which the echo is not faint at last.

Joachim du Bellay mentioned the herb in his "A Vow To Heavenly Venus," ca. 1500.

We that with like hearts love, we lovers twain,

New wedded in the village by thy fane,

Lady of all chaste love, to thee it is

We bring these amaranths, these white lilies,

A sign, and sacrifice; may Love, we pray,

Like amaranthine flowers, feel no decay;

Like these cool lilies may our loves remain,

Perfect and pure, and know not any stain;

And be our hearts, from this thy holy hour,

Bound each to each, like flower to wedded flower.

In ancient Greece, the amaranth (also called chrysanthemum and helichrysum) was sacred to Ephesian Artemis. It was supposed to have special healing properties, and, as a symbol of immortality, was used to decorate images of the gods and tombs. In legend, Amarynthus (a form of Amarantus) was a hunter of Artemis and king of Euboea; in a village of Amarynthus, of which he was the eponymous hero, there was a famous temple of Artemis Amarynthia or Amarysia (Strabo x. 448; Pausan. i. 31, p. 5). It was also widely used by the Chinese for its healing chemicals, curing illnesses such as infections, rashes, and migraines. The *"Amarantos"* is the name of a several-century-old popular Greek folk song:

Look at the amaranth:

on tall mountains it grows,

on the very stones and rocks

and places inaccessible.

Amaranth Grain

Amaranth has been cultivated as a grain for 8,000 years. The yield of grain amaranth is comparable to rice or maize. It was a staple food of the Aztecs, and was used as an integral part of Aztecreligious ceremonies.

The cultivation of amaranth was banned by the conquistadores upon their conquest of the Aztec nation. Because the plant has continued to grow as a weed since that time, its genetic base has been largely maintained. Research on grain amaranth began in the US in the 1970s. By the end of the 1970s, a few thousand acres were being cultivated. Much of the grain currently grown is sold in health food shops.

Grain amaranth is also grown as a food crop in limited amounts in Mexico, where it is used to make a candy called *alegría* (Spanish for happiness) at festival times. Amaranth species that are still used as a grain are: *Amaranthus caudatus*, *Amaranthus cruentus*, and *Amaranthus hypochondriacus*. The grain is popped and mixed with honey.

In Maharashtra state of India, it is called "Rajgira" in the Marathi language. The popped grain is mixed with melted jaggery in proper proportion to make iron and energy rich "laddus," a popular food provided at the Mid-day Meal Program in municipal schools.

Amaranth grain can also be used to extract amaranth oil- a particularly valued pressed seed oil with many commercial uses.

Nutritional Analysis

As the following table shows, in a raw form, grain amaranth has many nutrients.

Raw amaranth grain, however, isn't edible and can not be digested. Amaranth grain must be prepared and cooked like other grains. Another table below suggests cooked amaranth is a competing and promising source of nutrition when compared to wheat bread, higher in some nutrients and lower in others. Notable nutritional content attributes of raw amaranth grain include:-

- The protein is of an unusually high quality, according to ECHO.
- The actual nutritional value of amaranth as human food is less than would be expected from raw amaranth grain data. According to ECHO, this is due to anti-nutritional factors in raw amaranth grain; examples of anti-nutritional factors present in amaranth include oxalates, nitrates, saponins and

phenolic compounds. Cooking methods such as boiling amaranth in water and then discarding the water may reduce its toxic effects.

· A ¼ cup of raw amaranth grain supplies 60% of the Recommended Dietary Allowance of iron.

· Amaranth grain is particularly high in lysine, an amino acid that is low in other grains.,

· Amaranth grain is deficient in essential amino acids such as leucine and threonine,- both of which are present in wheat germ.,

· Amaranth grain is free of gluten, which is important for people with gluten allergies.

Synopsis~composition: Component (per 100g portion)	Amaranth Amount	Wheat Amount	Rice Amount	Sweetcorn Amount	Potato Amount
water (g)	11	11	12	76	82
energy (kJ)	1554	1506	1527	360	288
protein (g)	14	23	7	3	1.7
fat (g)	7	10	1	1	0.1
carbohydrates (g)	65	52	79	19	16
fibre (g)	7	13	1	3	2.4
sugars (g)	1.7	<0.1	>0.1	3	1.2
iron (mg)	7.6	6.3	0.8	0.5	0.5
manganese (mg)	3.4	13.3	1.1	0.2	0.1
calcium (mg)	159	39	28	2	9
magnesium (mg)	248	239	25	37	21
phosphorus (mg)	557	842	115	89	62
potassium (mg)	508	892	115	270	407
zinc (mg)	2.9	12.3	1.1	0.5	0.3
pantothenic acid (mg)	1.5	2.3	1.0	0.7	0.3
vitB6 (mg)	0.6	1.3	0.2	0.1	0.2
folate (µg)	82	281	8	42	18
thiamin (mg)	0.1	1.9	0.1	0.2	0.1
riboflavin (mg)	0.2	0.5	>0.1	0.1	>0.1
niacin (mg)	0.9	6.8	1.6	1.8	1.1

The table below presents nutritional values of cooked, edible form of amaranth grain to cooked, edible form of wheat grain as reported by United States Department of Agriculture's National Nutrient Database for Standard Reference, Release 23 (2010).

Synopsis~ composition: Component (per 100g portion)	Amaranth grain, cooked Amount	Bread, wheat germ Amount
water (g)	75	37
energy (kJ)	429	1092
protein (g)	4	10
fat (g)	2	3
carbohydrates (g)	19	48
fibre (g)	2	2
sugars (g)	n/a	4
iron (mg)	2.1	3.5
manganese (mg)	0.85	0.85
calcium (mg)	47	89
magnesium (mg)	65	28
phosphorus (mg)	148	121
potassium (mg)	135	254
zinc (mg)	0.9	1.0
pantothenic acid (mg)	<0.1	0.5
vitB6 (mg)	0.11	0.08
folate (µg)	22	118
thiamin (mg)	<0.1	0.4
riboflavin (mg)	0.02	0.38
niacin (mg)	0.24	4.5

Additional Agricultural Information

Amaranth from its start as a plant has literally a worldwide population currently where there are varieties for greens, varieties for grain, combinations and even ornamentals. The Great Plains has seen a surge in this crop from Rodale Farms developed varieties.

Amaranthus retroflexus, or pigweed, is a wild amaranth species in the United States. The name derives from the plant's tendency to sprout where hogs are pasture fed. Although both the leaves and seeds are edible, pigweed has not been cultivated as a food crop.

The virtue of amaranth is in light soils, it will produce food under harsh and lackluster nutrient conditions much like grain sorghum. It is a very efficient grain crop.

Nomenclature

Nomenclature is a term that applies to either a list of names and/or terms, or to the system of principles, procedures and terms related to naming- which is the assigning of a word or phrase to a particular object or property. The principles of naming vary from the relatively informal conventions of everyday speech to the internationally-agreed principles, rules and recommendations that govern the formation and use of the specialist terms used in scientific and other disciplines.

Naming "things" is a part of our general communication using words and language: it is an aspect of everyday taxonomy as we distinguish the objects of our experience, together with their similarities and differences, which we identify, name and classify. The use of names, as the many different kinds of nouns embedded in different languages, connects nomenclature to theoretical linguistics, while the way we mentally structure the world in relation to word meanings and experience relates to the philosophy of language.

Onomastics, the study of proper names and their origins, includes: anthroponymy, concerned with human names, including personal names, surnames and nicknames; toponymy the study of place names; and etymology, the derivation, history and use of names as revealed through comparative and descriptive linguistics.

The scientific need for simple, stable and internationally-accepted systems for naming objects of the natural world has generated many formal nomenclatural systems. Probably the best known of these nomenclatural systems are the five codes of biological nomenclature that govern the Latinized scientific names of organisms.

Definition & Criteria

Nomenclature is a system of words used in particular discipline. It is used in respect of giving names systematically following the rules to all known living.

Etymology

The word *nomenclature* is derived from the Latin *nomen-* name, *calare-* to call; the Ancient Greek from *onoma* meaning *name* and equivalent to the Old English *nama* and Old High German *namo* which is derived from Sanskrit *nama*. The Latin term *nomenclatura* refers to a list of names as does the word *nomenclator* which can also indicate a provider or announcer of names.

Onomastics and Nomenclature

The study of proper names is known as onomastics, which has a wide-ranging scope encompassing all names, all languages, all geographical and cultural regions. The distinction between onomastics and nomenclature is not readily clear: onomastics is an unfamiliar discipline to most people and the use of nomenclature in an academic sense is also not commonly known. Although the two fields integrate, nomenclature concerns itself more with the rules and conventions that are used for the formation of names.

Naming as a Cultural Activity

Names provide us with a way of structuring and mapping the world in our minds so, in some way, they mirror or represent the objects of our experience.

Names, Words, Language and Meaning

Elucidating the connections between language (especially names and nouns), meaning and the way we perceive the world has provided a rich field of study for philosophers and linguists. Relevant areas of study include: the distinction between proper names and proper nouns; and the relationship between names, their referents, meanings (semantics), and the structure of language.

Folk Taxonomy

Modern scientific taxonomy has been described as *"basically a Renaissance codification of folk taxonomic principles."* Formal scientific nomenclatural and classification systems are exemplified by biological classification. All classification systems are established for a purpose. The scientific classification system anchors each organism within the nested hierarchy of internationally-accepted classification categories. Maintenance of this system involves formal rules of nomenclature and periodic international meetings of review. This modern system evolved from the folk taxonomy of prehistory. Folk taxonomy can be illustrated through the Western tradition of horticulture and gardening. Unlike scientific taxonomy, folk taxonomies serve many purposes. Examples in horticulture would be the grouping of plants, and naming of these groups, according to their properties and uses: annuals, biennials and perennials (nature of life cycle); vegetables, fruits, culinary herbs and spices (culinary use); herbs, trees and shrubs (growth habit); wild and cultivated plants (whether they are managed or not), and weeds (whether they are considered to be a nuisance or not) and so on.

Folk taxonomy is generally associated with the way rural or indigenous peoples use language to make sense of and organise the objects around them. Ethnobiology frames this interpretation through either "utilitarianists" like Bronislaw Malinowski who maintain that names and classifications reflect mainly material concerns, and "intellectualists" like Claude Lévi-Strauss who hold that they spring from innate mental processes. The literature of ethnobiological classifications was reviewed in 2006. Folk classification is defined by the way in which members of a language community name and categorise plants and animals whereas ethnotaxonomy refers to the hierarchical structure, organic content, and cultural function of biological classification that ethnobiologists find in every society around the world.

Ethnographic studies of the naming and classification of animals and plants in non-Western societies have revealed some general principles that indicate pre-scientific man's conceptual and linguistic method of organising the biological world in a hierarchical way. Such studies indicate that the urge to classify is a basic human instinct.

- in all languages natural groups of organisms are distinguished (present-day taxa)
- these groups are arranged into more inclusive groups or ethnobiological categories
- in all languages there are about five or six ethnobiological categories of graded inclusiveness
- these groups (ethnobiological categories) are arranged hierarchically, generally into mutually exclusive ranks
- the ranks at which particular organisms are named and classified is often similar in different cultures

The levels are — moving from the most to least inclusive:

- level 1- "unique beginner" —e.g. *plant* or *animal*. A single all-inclusive name rarely used in folk taxonomies but loosely equivalent to an original living thing, a "common ancestor"
- level 2- "life form" ———————e.g. *tree, bird, grass* and *fish* These are usually primary lexemes (basic linguistic units) loosely equivalent to a phylum or major biological division.
- level 3- "generic name" ———e.g. *oak, pine, robin, catfish* This is the most numerous and basic building block of all folk taxonomies, the most frequently referred to, the most important psychologically, and among the first learned by children. These names can usually be associated directly with a second level group. Like life-form names these are primary lexemes.
- level 4- "specific name" ———e.g. *white fir, post oak* More or less equivalent to species. A secondary lexeme and generally less frequent than generic names.
- level 5- "varietal name"————e.g. *baby lima bean, butter lima bean.*

In almost all cultures objects are named using one or two words equivalent to "kind" (genus) and "particular kind" (species). When made up of two words (a binomial) the name usually consists of a noun (like *salt*, *dog* or *star*) and an adjectival second word that helps describe the first, and therefore makes the name, as a whole, more "specific", for example, *lap dog*, *sea salt*, or *film star*. The meaning of the noun used for a common name may have been lost or forgotten (*whelk*, *elm*, *lion*, *shark*, *pig*) but when the common name is extended to two or more words much more is conveyed about the organism's use, appearance or other special properties (*sting ray*, *poison apple*, *giant stinking hogweed*, *hammerhead shark*). These noun-adjective binomials are just like our own names with a family or surname like *Simpson* and another adjectival Christian- or forename name that specifies which Simpson, say *Homer Simpson*. It seems reasonable to assume that the form of scientific names we call binomial nomenclature is derived from this simple and practical way of constructing common names- but with the use of Latin as a universal language.

In keeping with the "utilitarianist" view other authors maintain that ethnotaxonomies resemble more a "complex web of resemblances" than a neat hierarchy.

Names and Nouns

A name is a label for any noun. Names can identify a class or category of things, or a single thing, either uniquely, or within a given context. Names ar given, for example, to humans or any other organisms, places, products- as in brandnames- and even to ideas or concepts. It is names as nouns that are the building blocks of nomenclature.

The word "name" is possibly derived from the Proto-Indo-European language hypothesised word *nomn*. The distinction between names and nouns, if made at all, is extremely subtle although clearly "noun" refers to names as lexical categories and their function within the context of language, rather that as "labels" for objects and properties.

Personal Names

Human personal names are presented, used and categorised in many ways depending on the language and culture. It is customary nowadays to use two-word names or binomials. In the Christian tradition the first name is given at baptism and referred to as the Christian name, but it is also known as the given name, forename or simply the first name. In England prior to the Norman invasion of 1066 small communities of Celts, Anglo-Saxons and Scandinavians all used single names, each person being identified by a single name as either a personal name or nickname. As the population increased, it gradually became necessary to identify people further – giving rise to names like John the butcher, Henry from Sutton, and Roger son of Richard ... which naturally evolved into John Butcher, Henry Sutton, and Roger Richardson. We now know this additional name variously as the second name, last name, family name, surnames or occasionally the byname, and this natural tendency was accelerated by the Norman tradition of using surnames that were fixed and hereditary within individual families. In combination these two names are now known as the personal name or, simply, the name. There are many exceptions to this general rule: Westerners often insert a third or more names between the given and surnames; Chinese and Hungarian names have the family name preceding the given name; females now often retain their maiden names (their family surname) or combine, using a hyphen, their maiden name and the surname of their husband; some East Slavic nations insert the patronym (a name derived from the given name of the father) between the given and the family name; in Iceland the given name is used with the patronym and surnames rarely used. Nicknames (sometimes called hypocoristic names) are informal names used mostly between friends.

Common Names and Proper Names

The distinction between proper names and common names is that proper names denote a unique entity e.g. London Bridge, while common names are used in a more general sense in reference to a class of objects e.g. bridge. Many proper names are obscure in meaning as they lack any apparent meaning in the way that ordinary words mean probably for the practical reason that when they consist of Collective nouns refer to groups, even when they are inflected for the singular e.g. "committee". Concrete nouns like "cabbage" refer to physical bodies that can be observed by at least one of the senses while abstract nouns, like "love" and "hate" refer to abstract objects. In English, many abstract nouns are formed by adding noun-forming suffixes ("-ness", "-ity", "-tion") to adjectives or verbs e.g. "happiness", "serenity" "concentration". Pronouns, like "he", "it", "which", and "those" stand in place of nouns in noun phrases.

The capitalisation of nouns varies with language and even the particular context: journals often have their own house styles for common names.

Onym Nouns

Distinctions may be made between particular kinds of names simply by using the suffix-onym, from the Greek Díìà (ónoma) 'name'. So we have, for example, hydronyms name bodies of water, synonyms are names with the same meaning and so on. Although the entire field could be described as chrematonymy- the names of things.

Toponyms

Toponyms are the names given to places or features of a particular district, region, etc. This could include planets, countries, cities, towns, villages, buildings etc.; it can be further divided into specialist branches: hodonymy, the names of streets, roads, and the like; hydronymy, the names of water bodies; and oronomy, the names of mountains. Toponymy has popular appeal because of its socio-cultural and historical interest and significance for cartography. However, work on the etymology of toponyms has found that many place names are descriptive, honourific or commemorative but frequently they have no meaning or the meaning is obscure or lost. Also the many categories of names are frequently interrelated. For example, many place-names are derived from personal names (Victoria), many names of planets and stars are derived from the names of mythological characters (Venus, Neptune), and many personal names are derived from place-names, names of nations and the like (Wood, Smith, Bridge).

Scientific Nomenclature

Nomenclature, Classification and Identification

In biological science, at least, nomenclature is regarded as a part of taxonomy. Taxonomy can be defined as the study of classification including its principles, procedures and rules, while classification itself is the ordering of taxa (the objects of classification) into groups based on similarities and/or differences. Doing taxonomy entails identifying, describing and naming taxa so nomenclature, in this strict scientific sense, is that branch of taxonomy concerned with the application of scientific names to taxa, based on a particular classification scheme and in accordance with agreed international rules and conventions.

Identification determines whether a particular taxon (singular of taxa) is identical to another one that has already been classified and named – so classification must precede identification. As taxa are rarely strictly identical, but only treated as such for the purposes of the classification, this procedure is sometimes referred to as "determination".

The precision demanded by science in the accurate naming of objects in the natural world has resulted in a variety of international nomenclatural codes, standards and protocols.

Biology

Although Linnaeus' system of binomial nomenclature was rapidly adopted after the publication of his Species Plantarum and Systema Naturae in 1753 and 1758 respectively, it was a long time before there was international consensus concerning the more general rules governing biological nomenclature.

The first botanical code was produced in 1905, the zoological code in 1889 and cultivated plant code in 1953.

Agreement on the nomenclature and symbols for genes emerged in 1979.

- *International Code of Botanical Nomenclature*, Botanical nomenclature
- *International Code of Nomenclature of Bacteria*
- *International Code of Nomenclature for Cultivated Plants*
- *International Code of Zoological Nomenclature*
- Virus nomenclature, used in Virus classification
- Enzyme nomenclature
- *PhyloCode* (the *International Code of Phylogenetic Nomenclature*, Phylogenetic nomenclature)- a new convention currently under development.
- *International standard on human anatomic terminology-Terminologia Anatomica*
- Gene nomenclature
- Red Cell Nomenclature
- Medical devices use the naming convention of the Global Medical Device Nomenclature (GMDN).

Astronomy

Over the last few hundred years, the number of identified astronomical objects has risen from hundreds to over a billion, and more are discovered every year. Astronomers need universal systematic designations to unambiguously identify all of these objects using astronomical naming conventions, while assigning names to the most interesting objects and, where relevant, naming important or interesting features of those objects.

- Planetary nomenclature
- Meteorite Nomenclature
- International Astronomical Union.

Chemistry

The IUPAC nomenclature is a system of naming chemical compounds and for describing the science of chemistry in general. It is maintained by the International Union of Pure and Applied Chemistry. The rules for naming organic and inorganic compounds are printed in two publications, the *Blue Book* and the *Red Book* available here. A third publication, *Green Book*, contains recommendations for the use of symbols for physical quantities (in association with the IUPAP), while a fourth, the *Gold Book*, defines a large number of technical terms used in chemistry. Similar compendia exist for biochemistry (in association with the IUBMB), analytical chemistry and macromolecular chemistry. These books are supplemented by shorter recommendations for specific circumstances which are published from time to time in the journal *Pure and Applied Chemistry*. These systems can be accessed here.

• International Union of Pure and Applied Chemistry (IUPAC).

Information Technology

There are hundreds of different nomenclatures used in the IT industry and the list is ever expanding due to technological advances and new innovations. Some of the more common nomenclatures used are: SoC(System on Chip), BIOS(Basic Input Output System), CPU(Central Processing Unit), and etc.

Metallurgy

The classic English translation of *De re metallica* includes an appendix (Appendix C) detailing problems of nomenclature in weights and measures.

Chapter 3

Controlled Vocabulary

Controlled vocabularies provide a way to organise knowledge for subsequent retrieval. They are used in subject indexing schemes, subject headings, thesauri, taxonomies and other form of knowledge organisation systems. Controlled vocabulary schemes mandate the use of predefined, authorised terms that have been preselected by the designer of the vocabulary, in contrast to natural language vocabularies, where there is no restriction on the vocabulary.

In Library and Information Science

In library and information science controlled vocabulary is a carefully selected list of words and phrases, which are used to tag units of information (document or work) so that they may be more easily retrieved by a search. Controlled vocabularies solve the problems of homographs, synonyms and polysemes by a bijection between concepts and authorised terms. In short, controlled vocabularies reduce ambiguity inherent in normal human languages where the same concept can be given different names and ensure consistency.

For example, in the Library of Congress Subject Headings (a subject heading system that uses a controlled vocabulary), authorised terms — subject headings in this case — have to be chosen to handle choices between variant spellings of the same concept (American versus British), choice among scientific and popular terms (Cockroaches versus Periplaneta americana), and choices between synonyms (automobile versus cars), among other difficult issues.

Choices of authorised terms are based on the principles of *user warrant* (what terms users are likely to use), *literary warrant* (what terms are generally used in the literature and documents), and *structural warrant* (terms chosen by considering the structure, scope of the controlled vocabulary).

Controlled vocabularies also typically handle the problem of homographs, with qualifiers. For example, the term "pool" has to be qualified to refer to either swimming pool, or the game pool to ensure that each authorised term or heading refers to only one concept.

There are two main kinds of controlled vocabulary tools used in libraries: subject headings and thesauri. While the differences between the two are diminishing, there are still some minor differences.

Historically subject headings were designed to describe books in library catalogues by catalogers while thesauri were used by indexers to apply index terms to documents and articles. Subject headings tend to be broader in scope describing whole books, while thesauri tend to be more specialised covering very specific disciplines. Also because of the card catalogue system, subject headings tend to have terms that are in indirect order (though with the rise of automated systems this is being removed), while thesaurus terms are always in direct order. Subject headings also tend to use more pre-coordination of terms such that the designer of the controlled vocabulary will combine various concepts together to form one authorised subject heading. (e.g., children and terrorism) while thesauri tend to use singular direct terms. Lastly thesauri list not only equivalent terms but also narrower, broader terms and related terms among various authorised and non-authorised terms, while historically most subject headings did not.

For example, the Library of Congress Subject Heading itself did not have much syndetic structure until 1943, and it was not until 1985 when it began to adopt the thesauri type term "Broader term" and "Narrow term".

The terms are chosen and organised by trained professionals (including librarians and information scientists) who possess expertise in the subject area. Controlled vocabulary terms can accurately describe what a given document is actually about, even if the terms themselves do not occur within the document's text. Well known subject heading systems include the Library of Congress system, MeSH, and Sears. Well known thesauri include the Art and Architecture Thesaurus and the ERIC Thesaurus.

Choosing authorised terms to be used is a tricky business, besides the areas already considered above, the designer has to consider the specificity of the term chosen, whether to use direct entry, inter consistency and stability of the language. Lastly the amount of pre-co-ordinate (in which case the degree of enumeration versus synthesis becomes an issue) and post co-ordinate in the system is another important issue.

Controlled vocabulary elements (terms/phrases) employed as tags, to aid in the content identification process of documents, or other information system entities (e.g. DBMS, Web Services) qualifies as metadata.

Indexing Languages

There are three main types of indexing languages.

- Controlled indexing language- Only approved terms can be used by the indexer to describe the document
- Natural language indexing language- Any term from the document in question can be used to describe the document.
- Free indexing language- Any term (not only from the document) can be used to describe the document.

When indexing a document, the indexer also has to choose the level of indexing exhaustivity, the level of detail in which the document is described. For example using low indexing exhaustivity, minor aspects of the work will not be described with index terms. In general the higher the indexing exhaustivity, the more terms indexed for each document.

In recent years free text search as a means of access to documents has become popular. This involves using natural language indexing with an indexing exhaustively set to maximum (every word in the text is *indexed*). Many studies have been done to compare the efficiency and effectiveness of free text searches against documents that have been indexed by experts using a few well chosen controlled vocabulary descriptors.

Controlled vocabularies are often claimed to improve the accuracy of free text searching, such as to reduce irrelevant items in the retrieval list. These irrelevant items (false positives) are often caused by the inherent ambiguity of natural language. Take the English word *football* for example. *Football* is the name given to a number of different team sports. Worldwide the most popular of these team sports is Association football, which also happens to be called *soccer* in several countries. The English language word football is also applied to Rugby football (Rugby union and rugby league), American football, Australian rules football, Gaelic football, and Canadian football. A search for *football* therefore will retrieve documents that are about several completely different sports. Controlled vocabulary solves this problem by tagging the documents in such a way that the ambiguities are eliminated.

Compared to free text searching, the use of a controlled vocabulary can dramatically increase the performance of an information retrieval system, if performance is measured by precision (the percentage of documents in the retrieval list that are actually relevant to the search topic).

In some cases controlled vocabulary can enhance recall as well, because unlike natural language schemes, once the correct authorised term is searched, you don't need to worry about searching for other terms that might be synonyms of that term.

However, a controlled vocabulary search may also lead to unsatisfactory recall, in that it will fail to retrieve some documents that are actually relevant to the search question.

This is particularly problematic when the search question involves terms that are sufficiently tangential to the subject area such that the indexer might have decided to tag it using a different term (but the searcher might consider the same). Essentially, this can be avoided only by an experienced user of controlled vocabulary whose understanding of the vocabulary coincides with the way it is used by the indexer.

Another possibility is that the article is just not tagged by the indexer because indexing exhaustivity is low. For example an article might mention football as a secondary focus, and the indexer might decide not to tag it with "football" because it is not important enough compared to the main focus. But it turns out that for the searcher that article is relevant and hence recall fails. A free text search would automatically pick up that article regardless.

On the other hand free text searches have high exhaustivity (you search on every word) so it has potential for high recall (assuming you solve the problems of synonyms by entering every combination) but will have much lower precision.

Controlled vocabularies are also quickly out-dated and in fast developing fields of knowledge, the authorised terms available might not be available if they are not updated regularly. Even in the best case scenario, controlled language is often not as specific as using the words of the text itself. Indexers trying to choose the appropriate index terms might misinterpret the author, while a free text search is in no danger of doing so, because it uses the author's own words.

The use of controlled vocabularies can be costly compared to free text searches because human experts or expensive automated systems are necessary to index each entry. Furthermore, the user has to be familiar with the controlled vocabulary scheme to make best use of the system. But as already mentioned, the control of synonyms, homographs can help increase precision.

Numerous methodologies have been developed to assist in the creation of controlled vocabularies, including faceted classification, which enables a given data record or document to be described in multiple ways.

Applications

Controlled vocabularies, such as the Library of Congress Subject Headings, are an essential component of bibliography, the study and classification of books. They were initially developed in library and information science. In the 1950s, government agencies began to develop controlled vocabularies for the burgeoning journal literature in specialised fields; an example is the Medical Subject Headings (MeSH) developed by the U.S. National Library of Medicine. Subsequently, for-profit firms (called Abstracting and indexing services) emerged to index the fast-growing literature in every field of knowledge. In the 1960s, an online bibliographic database industry developed based on dialup X.25 networking. These services were seldom made available to the public because they were difficult to use; specialist librarians called search intermediaries handled the searching job. In the 1980s, the first full text databases appeared; these databases contain the full text of the index articles as well as the bibliographic information. Online bibliographic databases have migrated to the Internet and are now publicly available; however, most are proprietary and can be expensive to use. Students enrolled in colleges and universities may be able to access some of these services without charge; some of these services may be accessible without charge at a public library.

In large organisations, controlled vocabularies may be introduced to improve technical communication. The use of controlled vocabulary ensures that everyone is using the same word to mean the same thing. This consistency of terms is one of the most important concepts in technical writing and knowledge management, where effort is expended to use the same word throughout a document or organisation instead of slightly different ones to refer to the same thing.

Web searching could be dramatically improved by the development of a controlled vocabulary for describing Web pages; the use of such a vocabulary could culminate in a Semantic Web, in which the content of Web pages is described using a machine-readable metadata scheme. One of the first proposals for such a scheme is the Dublin Core Initiative. An example of a controlled vocabulary which is usable for indexing web pages is PSH. It is unlikely that a single metadata scheme will ever succeed in describing the content of the entire Web.

To create a Semantic Web, it may be necessary to draw from two or more metadata systems to describe a Web page's. The exchangeable Faceted Metadata Language (XFML) is designed to enable controlled vocabulary creators to publish and share metadata systems. XFML is designed on faceted classification principles.

Senecio Vulgaris

Senecio vulgaris, often known by the common name Common groundsel, a humble in appearance member of the Asteraceae family and *Senecio* genus, is a tenacious deciduous annual whose presence now encompasses the globe in a wide area of easy to somewhat difficult growing conditions. The discussion of Common groundsel dates back to the 1st century and more recently, it is the subject of much contradictory and reactionary information about where it came from, how it got there, whether is it really hurting the crops, how to get rid of it, and how dangerous it is when ingested by various animals.

Description

This Senecio vulgaris is under a Sow Thistle.

Standing only between 4 and 16 inches (10 to 41 cm) tall, bright florets mostly hidden by the characteristic bract giving it the appearance of never opening flowers and with a life span of 5–6 weeks, the self fertilizing *Senecio vulgaris* lives humbly among and occasionally under the other weeds and is easy not to notice.

Close-up of Senecio vulgaris flowers.

Leaves and Stems

Leaves of *Senecio vulgaris* grow directly from the stem, sessile or lacking their own stem (petiole), alternating in direction along the length of the plant, two rounded lobes at the base of the stem (auriculate) and sub-clasping above. Leaves are pinnately lobed and +2.4 inches (61 mm) long and 1 inch (25 mm) wide and get smaller as up the plant. Leaves are covered sparsely with soft, smooth, fine hairs. Lobes typically sharp to rounded saw-toothed.

The hollow succulent stems branch at the tops and from the base. Stems and leaves can both host the Cinerarea leaf rust.

Flowers

Open clusters of 8 to 10 small cylinder shaped rayless yellow flower heads ¼ to ½ inch (6 to 13 mm) with a highly conspicuous ring of black tipped bracts at the base of the inflorescence as is characteristic of many members of the genus *Senecio*.

Seeds

The name for the genus *Senecio* is probably derived from Senex (an old man), in reference to its downy head of seeds; "the flower of this herb hath white hair and when the wind bloweth it away, then it appeareth like a bald-headed man" and like its family, flowers of *Senecio vulgaris* are succeeded by downy globed heads of seed. The seeds are achene, include a pappus and become sticky when wet. Laboratory tests have suggested maximum seed scattering distances of 2.1 and 3.2 yd (1.9 and 2.9 m) at wind speeds of 6.8 and 10.2 mph (10.9 and 16.4 km/h) respectively (affected by plant height) suggests that it was more than wind that spread these groundsel seeds throughout the world.

The average weight of 1000 seeds is 0.21 gram (2,200,000 seeds per pound) and experienced a 100% germination success before drying and storage and an 87% germination success after drying and 3 years of cool dry storage. In simple models for seed emergence prediction, soil thermal time did not predict the timing and extent of seedling emergence as well as hydrothermal time (warm rain).

Roots

The root system consists of a shallow taproot. This plant spreads by reseeding itself.

Groundsel acts as a host for the fungus that causes black root rot in peas, alfalfa, soybeans, carrots, tomatoes, red clover, peanuts, cucurbits, cotton, citrus,chickpeas, and several ornamental flowering plants; a list of flowering plants that can host their own fungus as well.

Common Names

English: *Old-man-in-the-Spring, Common Groundsel, Groundsel, Ragwort, Grimsel, Grinsel, Grundsel, Simson, Birdseed, Chickenweed, Old-man-of-the-spring, Squaw Weed, Grundy Swallow, Ground Glutton, Common Butterweed*:

- Portuguese: *Cardo-morto*
- Norwegian: *Åkersvineblom*
- Danish: *Almindelig Brandbæger*
- Croatian: *Badeljac, Gušèernjak, Kostriè zeèji, Obièni dragušac, Obièni kostriš, Obièni staraèac*
- German: *Gemeines Greiskraut, Gemeines Kreuzkraut, Gewöhnliches Greiskraut*
- Estonian: *Harilik ristirohi*
- Maltese: *&axixa tal-Kanali, Kubrita*
- French: *Herbe aux coitrons, Séneçon commun, Séneçon vulgaire*
- Galician: *Mexacán*
- Catalan: *Xenixell*
- Dutch: *Klein kruiskruid*

- Swedish: *Korsört, Vanlig korsört*
- Icelandic: *Krossgras*
- Slovene: *Navadni grint*
- Lithuanian: *Paprastoji þilë*
- Latvian: *Parastâ krustaine*
- Finnish: *Peltovillakko*
- Polish: *Starzec zwyczajny*
- Italian: *Senecione comune*
- Spanish: *Hierba cana, senecio común, flor amarilla, cineraria o yuyito*
- Slovak: *Starèek obyèajný*
- Romanian: *Cruciuliþã.*

Distribution

Senecio vulgaris is a frost resistant deciduous annual plant that grows willingly in disturbed sites, waste places, roadsides, gardens, nurseries, orchards, vineyards, landscaped areas, agricultural lands, at altitudes up to 1,600 feet (500 m) and is, additionally, self-pollinating producing 1,700 seeds per plant with three generations per year. Seeds are dispersed by wind and also cling to clothing and animal fur, and as contaminates of commercially exchanged seeds; the relocation of this plant throughout the planet has been difficult if not impossible to contain.

There is conflicting information about the native status of *Senecio vulgaris* in various locations. The United States Department of Agriculture (USDA), Natural Resources Conservation Service Plants Profile Database considers it to be native to all 50 of the United States of America, Canada, Greenland, Saint Pierre and Miquelon, the same USDA through the Germplasm Resources Information Network (GRIN) considers it to be native only to parts of Afro-Eurasia. The Integrated Taxonomic Information System Organisation (ITIS), a partnership between many United States federal government departments and agencies states that the species has been introduced to the 50 United States, and the online journal Flora of North America calls it "probably introduced" to areas north of Mexico. Individual research groups claim it is not native to areas they oversee: Florida, Washington, Wisconsin, Saskatchewan, British Columbia, Missouri. The United States Geological Survey reports that Common Groundsel is exotic to all 50 states and all Canadian provinces with the exception of Georgia, Kentucky, Massachusetts, and Labrador.

Native

to Algeria, Libya, Morocco, Tunisia, Georgia, Republic of Adygea, Karachay-Cherkess Republic, Kabardino-Balkaria Republic, Republic of North Ossetia-Alania, Republic of Ingushetia, Chechen Republic, Republic of Dagestan, Amur Oblast, Chukotka Autonomous Okrug, Jewish Autonomous Oblast, Kamchatka Oblast, Koryak Autonomous Okrug, Khabarovsk Krai, Magadan Oblast, Primorsky Krai, Sakha (Yakutia) Republic, Sakhalin Oblast, South Korea, North Korea, Denmark, Faroe Islands, Finland, Iceland, Ireland, Norway, Sweden, United Kingdom, Austria, Belgium, Czech Republic, Slovakia, Germany, Hungary, Netherlands, Poland, Switzerland, Belarus,Estonia, Latvia, Lithuania, Moldova, Russian Central Federal District, Russian Southern Federal District, Ukraine, Albania, Bulgaria, Greece, Italy, Romania, Bosnia and Herzegovina, Croatia, the Republic of Macedonia, Montenegro, Serbia, and Slovenia, France and Portugal.

Current

America

Canada: Alberta, British Columbia, Manitoba, New Brunswick, Newfoundland and Labrador, Northwest Territories and Nunavut (not collected North of the Hudson Bay), Nova Scotia, Ontario, Prince Edward Island, Quebec, Saskatchewan, Yukon.

Mexico: Aguascalientes, Baja California Norte, Chiapas, Coahuila, Distrito Federal, Hidalgo, Jalisco, Estado de México, Nuevo León, Puebla, Tlaxcala, Veracruz.

North America: United States of America, Greenland, Saint Pierre and Miquelon

South America: Argentina, Bolivia, Chile, Ecuador, Peru.

Africa

Northern Africa: Algeria, Egypt including Sinai, Libya, Morocco and Spanish Morocco, Tunisia.

Asia

Western Asia: Iran, Israel with the Palestinian territories, Lebanon and Syria.

Caucasus: Adygea, Azerbaijan, Chechnya, Dagestan, Georgia, Ingushetia, Kabardino-Balkaria, Kalmykia, Karachay-Cherkessia, North Ossetia-Alania.

Northwestern Asia: Arkhangelsk Oblast including Nenets Autonomous Okrug and Novaya Zemlya, Astrakhan Oblast, Bashkortostan, Belgorod Oblast, Bryansk Oblast, Chuvashia, Lipetsk Oblast, Kaliningrad Oblast, Kaluga Oblast, Novgorod Oblast, Republic of Karelia, Kirov Oblast, Komi Republic, Kursk Oblast, Mordovia, Murmansk Oblast, Orenburg Oblast, Penza Oblast, Perm Krai, Pskov Oblast, Rostov Oblast, Ryazan Oblast, Saint Petersburg, Samara Oblast, Saratov Oblast, Tambov Oblast, Tatarstan, Tula Oblast, Udmurtia, Volgograd Oblast, Vologda Oblast, Voronezh Oblast,

Siberia: Altai Krai, Buryatia, Chelyabinsk Oblast, Chita Oblast, Irkutsk Oblast, Kemerovo Oblast, Krasnoyarsk Krai, parts of Kurgan Oblast, Novosibirsk Oblast, Omsk Oblast, Sverdlovsk Oblast, Tomsk Oblast, Tuva, Tyumen Oblast, Ulyanovsk Oblast, Yamal-Nenets,

Soviet Far East: Amur Oblast, Chukotka Autonomous Okrug, Jewish Autonomous Oblast, Kamchatka Oblast, Koryak Autonomous Okrug, Khabarovsk Krai, Magadan Oblast, Primorsky Krai, Sakha (Yakutia) Republic, Sakhalin Oblast

China: Inner Mongolia, Heilongjiang, Jilin, Liaoning, Hebei and Beijing

Eastern Asia: South Korea, North Korea, Hokkaidô, Honshû, Shikoku, Kyûshû, Okinawa Island.

Europe

Northern Europe: Channel Islands, Denmark, Estonia, Faeroe Islands, Finland, Iceland, Ireland with Northern Ireland, Norway, Sweden, United Kingdom.

Middle Europe: Austria, Czech Republic, Germany, Poland, Slovakia, Slovenia,

East Europe: Belarus, Hungary, Latvia, Lithuania, Moldavia, Ukraine.

Southeastern Europe: Bosnia and Herzegovina, Bulgaria, Croatia, Republic of Macedonia, Montenegro, Romania, Serbia.

Southwestern Europe: Andorra, Azores, Balearic Islands, Canary Islands, Gibraltar, Portugal, Spain.

West Europe: Belgium, France, Liechtenstein, Luxembourg, Netherlands, Switzerland,

South Europe: Albania, Chios, Corsica, Crete, Cyprus, Dodecanese, Greece, Ikaria, Italy, Karpathos, Lesbos, Samos Island, San Marino, Sardinia, Sicily with Malta, Asiatic Turkey and Turkey-in-Europe, Vatican City.

Oceania

Australasia: Jarrah Forest, Swan Coastal Plain, Warren, New Zealand.

South America

Predators

Flame Shoulder moth orOchropleura plecta.

Cinnabar moth (Tyria jacobaeae) caterpillar feeding on a Senecio.

The seed of Common groundsel is a good green food for Canaries and Finches and it is available all year round.

Senecio vulgaris seed has been found in the droppings of sparrows, and seedlings have been raised from the excreta of various birds. Seed has also been found in cow manure.

Some *Lepidoptera* species eat many of the *Senecio* genus; additional studies via electrophysiological recordings have shown that the taste sensilla of the Cinnabar moth larvae respond (get excited) specifically to the pyrrolizidine alkaloids, which all *Senecio* are known to contain.

Moths and Caterpillars
- Cinnabar moth (*Tyria jacobaeae*)
- Flame Shoulder (*Ochropleura plecta*)
- Ragwort plume moth (*Platyptilia isodactyla*).

The *Senecio* genus also are host to other insects:

Beetles
- Ragwort flea beetle (*Longitarsus jacobaeae*)
- *Longitarsus gracilis* (family Coleoptera species *Chrysomelidae*)

Flies
Seed flies (Diptera: Muscoidea).
- Ragwort seed fly (Anthomyiidae, *Botanophila seneciella*)

Gall flies (Diptera: Tephritidae):
- *Ensina sonchi*
- *Sphenella marginata*
- *Trupanea stellata*
- *Trypeta zoe.*

The ragwort flea beetle and ragwort seed fly have been approved and released for Senecio control in California, Australia and elsewhere.

Fungus Most *Senecio*, including *S. squalidus* are susceptible to rust and other fungus and mildews:

Rust fungus Uredinales:
- *Coleosporium tussilaginis-* stems and leaves (Coleosporiaceae)
- *Puccinia lagenophorae-* leaves only (Pucciniaceae)
- *Bremia lactucae.*

White rust Peronosporales:
- *Albugo tragopogonis-* (Albuginaceae)
- some of the species *Peronosporaceae-* (Albuginaceae).

Sac fungus Ascochyta, Pezizomycetes:
- *Ascochyta senecionicola-* (Coelomycete).

Groundsel Mildew Erysiphales:
- *Golovinomyces cichoracearum* var. *fischeri.*

Powdery Mildew Erysiphales:

* *Podosphaera fusca-* (Erysiphaceae).

Black root rot Microascales:

* some of the family Incertae sedis.

Reputation for Being Noxious and Toxic

Senecio vulgaris has been listed as a noxious weed, being both non-indigenous to most if not all of the Americas and having a reputation for being hepatotoxic to livestock and to humans.

Toxic Versus Medicinal

Human

As a plant that is reported to be both poisonous for human ingestion and also medicinal; much of the contradiction can be found by closely reviewing the words that are used and the dose (amount) of the poisonous substance that is ingested to prove either claim. All species of the genus *Senecio* contain pyrrolizidine alkaloids (e.g., senecionine) a substance that when a human has *chronic exposure* can cause irreversible liver damage. Common groundsel as a medicinal herb does not seem to be recommended very often since 1931, when it was recommended as a diaphoretic, an antiscorbutic, a purgative, a diuretic and an anthelmintic, which was a demotion as it was previously suggested for the expelling of gravel of the kidneys and reins by Pedanius Dioscorides in the 70s-90s, for use as poultices by John Gerard in the late 16th century and as a cure for epilepsy by Nicholas Culpeper in the 17th century. More current information is contradictory about the dangers of the ingestion of Groundsel. A heavily referenced paper from 1989 suggests that the response is immediate and gives pre-ambulatory care recommendations. A Canadian poisonous plants information database references a paper from 1990 in presenting this prenatal warning: "In a case of prenatal exposure, a mother ingested tea containing an estimated 0.343 milligram of senecionine, resulting in fatal veno-occlusive disease in a newborn infant." Information about the pyrrolizidine alkaloids, the substance present in *Senecio vulgaris* is much less contradictory and all warn of accumulation of the alkaloid. Certain pyrrolizidine alkaloids are non-toxic precursors that are converted to toxic metabolites in the body in a process called toxification

Livestock

Linnaeus is cited to have claimed that "goats and swine eat this common plant freely, cows being not partial to it and horses and sheep declining to touch it, but not only are caged birds fond of it (the seeds), but its leaves and seeds afford food for many of our wild species (rabbits were given as an example)." More recent studies claim that the lethal amount that cattle or horses need to consume is 7% of their body weight (example: 50 pounds (23 kg) would need to be consumed by a cow weighing 700 pounds (318 kg)). Lesser amounts cause the liver to lose function but is not apparent until the animal is stressed (by new feed or location, pregnancy, a different toxin, etc.). Sheep and goats have rumenbacteria that detoxify the alkaloids and are able to consume twice their body weight of this and other species of genus *Senecio*. The alkaloids responsible are not destroyed by drying or by fermentation in silage.

Introduced Versus Invasive

Introduced species become invasive when they compete with natives or with crops. *Senecio vulgaris* is not known to be a strong competitor but it has been known to reduce mint production. There is evidence that it is not a strong invasive and sometimes protective of critically endangered native plants.

Control

The approximately 22 millimetres (0.87 in) long pappus seeds of *Senecio vulgaris*, each plant capable of producing 25,000 or more seeds (1,700 seeds per plant are more likely) with three generations of the plant per year; seeds that are widely dispersed by the wind, have been identified as a contaminant of cereal and vegetable seeds and a poison to some livestock; there is some inspiration to understand the growth stages and determine some control methods.

Cultivation: Cultivation with the hand or tiller is a recommended method of controlling *Senecio vulgaris* from growing in gardens and planting fields; cultivate to a depth of 2 inches (51 mm). The plant does prefer to take root in disturbed soils, so cultivation rids new plants but also buries and stirs up new seeds so the cultivation needs to be repeated at 14-day intervals. Seeds can still mature even when the plant has been killed; seed from plants cut in flower had germination levels of 35%. Groundsel seed numbers increased in soil during a 2-year set-aside left fallow but not when there was a sown grass cover. The weed cannot live on grazed, trampled or mowed sites.

Biological: The pathogen rust fungus or *Puccinia lagenophorae* and the Cinnabar moth *Tyria jacobaeae* have both been used and studied in an attempt to control infestation of *Senecio vulgaris*. One study showed that rust fungus infected *Senecio vulgaris* survived and actually used more of the available soil nutrients. The cinnabar moth eats groundsel between June and August, but the seeds germinate and the plant grows as soon as the ground is warm enough (and after a warm rain), making this an insufficient control almost everywhere groundsel can be found.

Chemical: Herbicides designed to control broadleaf plants are affective for controlling *Senecio vulgaris* in cereals and forage grasses but also will "control" broadleaf crops, such as mint, forage legumes, strawberries, carrots and all other non-grass crops. There is also evidence that the plant develops an immunity to the chemical control.

Other: Groundsel seedlings with 2-6 leaves are tolerant of flame weeding but the seeds are susceptible to soil solarization.

Chapter 4

Environmental Protection

Environmental protection is a practice of protecting the environment, on individual, organisational or governmental level, for the benefit of the natural environment and (or) humans. Due to the pressures of population and our technology the biophysical environment is being degraded, sometimes permanently. This has been recognised and governments began placing restraints on activities that causedenvironmental degradation. Since the 1960s activism by the environmental movement has created awareness of the various environmental issues. There is not a full agreement on the extent of the environmental impact of human activity and protection measures are occasionally criticized.

Academic institutions now offer courses such as environmental studies, environmental management and environmental engineering that study the history and methods of environmental protection. Protection of the environment is needed from various human activities. Waste, pollution, loss of biodiversity, and the introduction of invasive species are some of the issues relating to environmental protection.

Evolving Approaches to Environmental Protection

Discussion concerning environmental protection often focuses on the role of government, legislation and enforcement, however in its broadest sense environmental protection may be seen to be the responsibility of all people and not simply that of government. Decisions that impact on the environment will ideally involve a broad range of stakeholders including industry, indigenous groups, environmental group and community representatives. Gradually environmental decision-making processes are evolving to reflect this broad base of stakeholders and are becoming more collaborative in many countries.

Environmental protection is influenced by three interwoven factors: environmental legislation, ethics and education. Each of these factors plays its part in influencing national level environmental decisions and personal level environmental values and behaviours. For environmental protection to become a reality it will be important for societies to develop each of these areas that together will inform and drive environmental decisions. Although environmental protection is not simply the role of government agencies they are however generally seen as being of prime importance in establishing and maintaining basic standards that protect both the environment and the people interacting with it. Outlined below are several approaches to environmental protection that are currently evolving. Further discussion on approaches to environmental protection is included on the pages related tonatural resource management, environmental governance and environmental law.

Voluntary Environmental Agreements

In industrialised countries voluntary environmental agreements often provide a platform for companies to be recognised for moving beyond the minimum regulatory standards and thus support the development of best environmental practice. In developing countries such as throughout Latin America, these agreements are more commonly used to remedy significant levels of non-compliance with mandatory regulation. The challenges that exist with these agreements lie in establishing baseline data, targets, monitoring and reporting. Due to the difficulties inherent in evaluating effectiveness their use is often questioned and indeed the environment may well be adversely affected as a result. The key advantage of their use in developing countries is that their use helps to build environmental management capacity.

Ecosystems Approach

An ecosystems approach to resource management and environmental protection aims to consider the complex interrelationships of an entire ecosystem in decision making rather than simply responding to specific issues and challenges. Ideally the decision-making processes under such an approach would be a collaborative approach to planning and decision-making that involves a broad range of stakeholders across all relevant government departments as well as representatives of industry, environmental groups and community. This approach ideally supports better exchange of information, development of conflict resolution strategies and improved regional conservation.

International Environmental Agreements

Many of the earth's resources are especially vulnerable because they are influenced by human impacts across many countries. As a result of this many attempts are made by countries to develop agreements that are signed by multiple governments to prevent damage or manage the impacts of human activity on natural resources. This can include agreements that impact on factors such as climate, oceans, rivers and air pollution. These international environmental agreements are sometimes legally binding documents that have legal implications when they are not followed and at other times are more agreements in principle or as codes of conduct. These agreements have a long history with some multinational agreements being in place from as early as 1910 in Europe, America and Africa. Some of the most well known multinational agreements include: the Kyoto Protocol, Vienna Convention on the Protection of the Ozone Layer and Rio Declaration on Development and Environment.

Government

Many Constitutions acknowledge the fundamental right to environmental protection and many international treaties acknowledge the right to live in a healthy environment.

But complete environmental protection seems impossible at this current global position.

Also, many countries have organisations and agencies devoted to environmental protection. There are International environmental protection organisations, as the United Nations Environment Programme.

Africa

Tanzania

Zebras, Serengeti savana plains, Tanzania

Tanzania is recognised as having some of the greatest biodiversity of any African country. Almost 40% of the land has been established into a network of protected area, including several national parks.

The concerns for the natural environment include damage to ecosystems and loss of habitat resulting from population growth, expansion of subsistence agriculture, pollution, timber extraction and significant use of timber as fuel.

History of Environmental Protection

Environmental protection in Tanzania began during the period of German occupation of East Africa (1884-1919) when colonial conservation laws for the protection of game and forests were enacted that placed restrictions on traditional indigenous activities such as hunting, collection of firewood and grazing of cattle. In 1948 Serengeti was officially established as the first national park in East Africa. Since 1983 there has been a more broad reaching effort to manage environmental issues at a national level through the establishment of the National Environment Management Council (NEMC) and development of an environmental act.

Government Protection

The Division of the Environment is the main government body that oversees protection. It does this through formulation of policy, co-ordination and monitoring of environmental issues, environmental planning and policy oriented environmental research. The National Environment Management Council (NEMC) is an institution that was initiated when the National Environment Management Act was first introduced in 1983. This council has the role to advise governments and the international community on a range of environmental issues. The NEMC has the following purposes: provide technical advice; co-ordinate technical activities; develop of enforcement guidelines and procedures; assess, monitor and evaluate activities that impact the environment; promote and assist environmental information and communication; seek advancement of scientific knowledge.

The National Environment Policy of 1997 acts as a framework for environmental decision-making in Tanzania. The policy objectives are to:

- Ensure sustainable and equitable use of resources without degrading the environment or risking health or safety
- Prevent and control degradation of land, water, vegetation and air
- Conserve and enhance natural and man-made heritage, including biological diversity of unique ecosystems
- Improve condition and productivity of degraded areas

- Raise awareness and understanding of linkages between environment and development
- Promote individual and community participation
- Promote international cooperation.

Tanzania is a signatory to a significant number of International Conventions including the Rio Declaration on Development and Environment 1992 and the Convention on Biological Diversity 1996. The Environmental Management Act, 2004 is the first comprehensive legal and institutional frameworks to guide environmental management decision. The policy tools that are part of the Act include the use of: environmental impact assessments; strategic environmental assessments and; taxation on pollution for specific industries and products. The effectiveness of this act will only become clear over time as concerns regarding its implementation are apparent based on the fact that historically there has been a lack of capacity to enforce environmental laws and a lack of working tools to bring environmental protection objectives into practice.

Asia

China

Formal environmental protection in China was first stimulated by the 1972 United Nations Conference on the Human Environment, held in Stockholm, Sweden. Following this, China began establishing environmental protection agencies and putting controls on some of its industrial waste. China was one of the first developing countries to implement a sustainable development strategy. In 1983 the State Council announced that environmental protection would be one of China's basic national policies and in 1984 the National Environmental Protection Agency (NEPA) was established. Following severe flooding of the Yangtze River basin in 1998, NEPA was upgraded to the State Environmental Protection Agency (SEPA) meaning that environmental protection was now being implemented at a ministerial level. In 2008, SEPA became known by its current name of Ministry of Environmental Protection of the People's Republic of China (MEP).

Pollution Control Instruments in China

Environmental pollution and ecological degradation has resulted in economic losses for China. In 2005, economic losses (mainly from air pollution) were calculated at 7.7% of China's GDP. This grew to 10.3% by 2002 and the economic loss from water pollution (6.1%) began to exceed that caused by air pollution. China has been one of the top performing countries in terms of GDP growth (9.64% in the past ten years). However, the high economic growth has put immense pressure on its environment and the environmental challenges that China faces are greater than most countries. In 2010 China was ranked 121st out of 163 countries on the Environmental Performance Index.

China has taken initiatives to increase its protection of the environment and combat environmental degradation:

- China's investment in renewable energy grew 18% in 2007 to $15.6 billion, accounting for ~10% of the global investment in this area;).

- In 2008, spending on the environment was 1.49% of GDP, up 3.4 times from 2000;

- The discharge of COD (carbon monoxide) and SO_2 (sulfur dioxide) decreased by 6.61% and 8.95% in 2008 compared with that in 2005;

- China's protected nature reserves have increased substantially. In 1978 there were only 34 compared with 2,538 in 2010. The protected nature reserve system now occupies 15.5% of the country; this is higher than the world average.

Rapid growth in GDP has been China's main goal during the past three decades with a dominant development model of inefficient resource use and high pollution to achieve high GDP. For China to develop sustainably, environmental protection should be treated as an integral part of its economic policies.

Quote from Shengxian Zhou, head of MEP (2009): "Good economic policy is good environmental policy and the nature of environmental problem is the economic structure, production form and develop model."

European Union

Environmental protection has become an important task for the institutions of the European Community after the Maastricht Treaty for the European Union ratification by all Member States. The EU is already very active in the field of environmental policy with important directives like those on environmental impact assessment and on the access to environmental information for citizens in the Member States.

Latin America

The United Nations Environment Programme (UNEP) has identified 17 megadiverse countries. The list includes six Latin American countries: Brazil, Colombia, Ecuador, Mexico, Peru and Venezuela. Mexico and Brazil stand out among the rest because they have the largest area, population and number of species. These countries represent a major concern for environmental protection because they have high rates of deforestation, ecosystems loss, pollution, and population growth.

Brazil

Brazil has the largest amount of the world's tropical forests, 4,105,401 km2 (48.1 % of Brazil), concentrated in the Amazon region. Brazil is home to vast biological diversity, first among the megadiverse countries of the world, having between 15%-20% of the 1.5 million globally described species.

The organisation in charge of environment protection is the Brazilian Ministry of the Environment (in Portuguese: Ministério do MeioAmbiente, MMA). It was first created in 1973 with the name Special Secretariat for the Environment (Secretaria Especial de Meio Ambiente), changing names several times, and adopting the final name in 1999. The Ministry is responsible for addressing the following issues:

- A national policy for the environment and for water resources;
- A policy for the preservation, conservation and sustainable use of ecosystems, biodiversity and forests;
- Proposing strategies, mechanisms, economic and social instruments for improving environmental quality, and sustainable use of natural resources;
- Policies for integrating production and the environment;
- Environmental policies and programs for the Legal Amazon;
- Ecological and economic territorial zoning.

In 2011, protected areas of the Amazon covered 2,197,485 km^2 (an area larger than Greenland), with conservation units, like national parks, accounting for just over half (50.6%), and indigenous territories representing the remaining 49.4%.

Mexico

With over 200,000 different species, Mexico is home to 10–12% of the world's biodiversity, ranking first in reptile biodiversity, second in mammals and second in diversity of ecosystems.

The history of environmental policy in Mexico started in the 1940s with the enactment of the Law of Conservation of Soil and Water (in Spanish: Ley de Conservación de Suelo y Agua). Three decades later, at the beginning of the 1970s, the Law to Prevent and Control Environmental Pollution was created (Ley para Preveniry

Controlar la Contaminación Ambiental). In 1972 was the first direct response from the federal government to address eminent health effects from environmental issues. It established the administrative organisation of the Secretariat for the Improvement of the Environment (Subsecretaría para el Mejoramiento del Ambiente) in the Department of Health and Welfare.

The axolotl is an endemic species from the central part of Mexico

The Secretariat of Environment and Natural Resources (Secretaría del Medio Ambiente y Recursos Naturales, SEMARNAT) is Mexico's environment ministry. The Ministry is responsible for addressing the following issues:

- Promote the protection, restoration and conservation of ecosystems, natural resources, goods and environmental services, and to facilitate their use and sustainable development.

- Develop and implement a national policy on natural resources

- Promote environmental management within the national territory, in coordination with all levels of government and the private sector.

- Evaluate and provide determination to the environmental impact statements for development projects and prevention of ecological damage

- Implement national policies on climate change and protection of the ozone layer.

- Direct work and studies on national meteorological, climatological, hydrological, and geohydrological systems, and participate in international conventions on these subjects.

- Regulate and monitor the conservation of waterways

In November 2000 there were 127 protected areas; currently there are 174, covering an area of 25,384,818 hectares, increasing federally protected areas from 8.6% to 12.85% its land area.

Oceania

Australia

In 2008 there was 98,487,116 ha of terrestrial protected area, covering 12.8% of the land area of Australia. The 2002 figures of 10.1% of terrestrial area and 64,615,554 ha of protected marine area were found to poorly represent about half of Australia's 85 bioregions.

Environmental protection in Australia could be seen as starting with the formation of the first National Park, Royal National Park, in 1879. More progressive environmental protection had it start in the 1960s and 1970s with major international programs such as the United Nations Conference on the Human Environment in 1972, the Environment Committee of the OECD in 1970, and the United Nations Environment Programme of 1972. These events laid the foundations by increasing public awareness and support for regulation. State environmental legislation was irregular and deficient until the Australian Environment Council (AEC) and Council of Nature Conservation Ministers (CONCOM) were established in 1972 and 1974, creating a forum to assist in coordinating environmental and conservation policies between states and neighbouring countries. These councils have since been replaced by the Australian and New Zealand Environment and Conservation Council (ANZECC) in 1991 and finally the Environment Protection and Heritage Council (EPHC) in 2001.

At a national level, the Environment Protection and Biodiversity Conservation Act of 1999 is the primary environmental protection legislation for the Commonwealth of Australia. It concerns matters of national and international environmental significance regarding flora, fauna, ecological communities and cultural heritage. It also has jurisdiction over any activity conducted by the Commonwealth, or affecting it, that has significant environmental impact. The act covers eight main areas:

- National Heritage Sites
- World Heritage Sites
- RAMSAR wetlands
- Nationally endangered or threatened species and ecological communities

- Nuclear activities and actions
- The Great Barrier Reef Marine Park
- Migratory species
- Commonwealth marine areas.

There are several Commonwealth protected lands due to partnerships with traditional native owners, such as Kakadu National Park, extraordinary biodiversity such as Christmas Island National Park, or managed cooperatively due to cross-state location, such as the Australian Alps National parks.

At a state level, the bulk of environmental protection issues are left to the responsibility of the state or territory. Each state in Australia has its own environmental protection legislation and corresponding agencies. Their jurisdiction is similar and covers point-source pollution, such as from industry or commercial activities, land/water use, and waste management. Most protected lands are managed by states and territories with state legislative acts creating different degrees and definitions of protected areas such as wilderness, national land and marine parks, state forests, and conservation areas. States also create regulation to limit and provide general protection from air, water, and sound pollution. At a local level, each city or regional council has responsibility over issues not covered by state or national legislation. This includes non-point source, or diffuse pollution, such as sediment pollution from construction sites.

Australia ranks second place on the UN 2010 Human Development Index and one of the lowest debt to GDP ratios of the developed economies. This could be seen as coming at the cost of the environment, with Australia being the world leader in coal exportation and species extinctions. Some have been motivated to proclaim it is Australia's responsibility to set the example of environmental reform for the rest of the world to follow.

New Zealand

At a national level the Ministry for the Environment is responsible for environmental policy and the Department of Conservation addresses conservation issues. At a regional level the regional councils administer the legislation and address regional environmental issues.

North America

United States

Since 1970, the United States Environmental Protection Agency (EPA) has been working to protect the environment and human health.

All U.S. states have their own state departments of environmental protection.

The EPA has drafted "Seven Priorities for EPA's Future", which are:

- "Taking Action on Climate Change"
- "Improving Air Quality"
- "Assuring the Safety of Chemicals"
- "Cleaning Up Our Communities"
- "Protecting America's Waters"
- "Expanding the Conversation on Environmentalism and Working for Environmental Justice"
- "Building Strong State and Tribal Partnerships".

In Literature

There are many works of literature that contain themes of environmental protection but some have been fundamental to its evolution. Several pieces such as *A Sand County Almanac* by Aldo Leopold, *Tragedy of the commons* by Garrett Hardin, and *Silent Spring* by Rachel Carson have become classics due to their far reaching influences.

Environmental protection is present in fiction as well as non-fictional literature. Books such as *Antarctica* and *Blockade* have environmental protection as subjects whereas *The Lorax* has become a popular metaphor for environmental protection. "The Limits of Trooghaft" by Desmond Stewart is a short story that provides insight in to human attitudes towards animals. Another book called "The Martian Chronicles" by Ray Bradbury investigates issues such as bombs, wars, government control, and what effects these can have on the environment.

Challenges in Environmental Protection

- The main issues for developing countries like Brazil and Mexico are that protected areas suffer from encroachment and poor management. In Brazil, protected areas are increasing but there is significant challenges caused by human impacts. Logging and mining are potentially huge threats to protected areas. Between 1998 and 2009, 12,204 km^2 of forest within protected areas was cleared, with 1,338 mining titles being granted and 10,348 awaiting approval. Developing countries need to allocate more money from their budgets if they hope to address these problems in environmental protection.

- Several challenges face African governments in implementing any environmental protection mechanisms. In Tanzania for example the challenges include: lack of financial resources to manage protected areas, poor governance and corruption and significant illegal logging and hunting. Also with such large allocations of land to national parks has been the displacement of indigenous people and a lack of local participation in environmental decision making processes. As a result of these factors recent calls have been made to allow "parks with people" as one means to support better overall management and care of the land.

- Due to the Australian climate being dominated by desert and semi-arid regions, most of the environmental protection challenges focus on availability and management of water resources. Even though this will continue to be an issue in areas of great demand, such as the Murray-Darling basin, several events were pivotal battles in environmental protection.

Case Study, Franklin River Dam:

In 1979, the building of an hydroelectric dam was proposed on the Franklin River in Western Tasmania. The advantages of this project would be increased power production and the creation of job in a region with one of the highest unemployment rates in Tasmania. Conservationist were concerned about the high concentration of Aboriginal sites and that it was one of Australia's last true wild rivers. The issue quickly became a focus of environmental protection, with the Tasmanian Wilderness Society leading the resistance movement. The situation escalated from a state referendum to a public blockade of construction, eventually leading to federal legislative intervention and a state challenge in the High Court. The state lost the case with the area proclaimed the Franklin-Gordon Wild Rivers National Park in 1981, part of the Tasmanian Wilderness World Heritage Area.

Conservation Movement

The conservation movement, also known as nature conservation, is a political, environmental and a social movement that seeks to protect natural resources including animal, fungus and plant species as well as their habitat for the future.

The early conservation movement included fisheries and wildlife management, water, soil conservation and sustainable forestry. The contemporary conservation movement has broadened from the early movement's emphasis on use of sustainable yield of natural resources and preservation of wilderness areas to include preservation of biodiversity. Some say the conservation movement is part of the broader and more far-reaching environmental movement, while others argue that they differ both in ideology and practice. Chiefly in the United States, conservation is seen as differing from environmentalism in that it aims to preserve natural resources expressly for their continued sustainable use by humans. In other parts of the world conservation is used more broadly to include the setting aside of natural areas and the active protection of wildlife for their inherent value, as much as for any value they may have for humans.

History

Jones (1991) argues that from an economic perspective the Western nations have been no more destructive of natural resources than any other civilization. He rejects the suggestion that Christianity, by destroying animism, facilitated the ruination of nature in the West, stating he finds no evidence that any culture was or is less exploitive of the natural world than Christianity. He notes that Eastern agricultural history has numerous examples of massive deforestations, erosion, silting of rivers, and infestation with waterborne parasites. He points to large-scale animal extinction and wasteful agricultural practices by North American Indians before 1492. Jones allows that economic growth in the West did result in a higher level of resource use, but finds no evidence to support the view that such resource exploitation was a product of religion, culture, or geography.

History of Conservation Ideas

The nascent conservation movement slowly developed in the 19th century, starting first in the scientific forestry methods pioneered by Prussia and France in the 17th and 18th centuries. While continental Europe created the scientific methods later used in conservationist efforts, British India and the United States are credited with starting the conservation movement.

Foresters in India, often German, managed forests using early climate change theories (in America,, George Perkins Marsh) that Alexander von Humboldt developed in the mid 19th century, applied fire protection, and tried to keep the "house-hold" of nature. This was an early ecological idea, in order to preserve the growth of delicate teak trees. The same German foresters who headed the Forest Service of India, such as Dietrich Brandis and Berthold Ribbentrop, travelled back to Europe and taught at forestry schools in England (Cooper's Hill, later moved to Oxford). These men brought with them the legislative and scientific knowledge of conservationism in British India back to Europe, where they distributed it to men such as Gifford Pinchot, which in turn helped bring European and British Indian methods to the United States.

Asia

India

Sivaramakrishnan (2009) explores the boundaries between wildness and civility in Indian society, as well as connection of ideas of nature to different aspects of social life, especially labour, aesthetics, politics, commerce, and agriculture.

These interconnected historical processes inform environmental history in India. At present forest history is the area of environmental history in which the most important scholarly debate is underway in India, with special interest in questions of water, air, industry, and climate change At the grass root level are organising mass movements with the theme of Think Globally-Act locally for conservation of nature since 1993 by Vijaypal baghel, peoples are called him ecoman, greenman etc.

So many events are conducting as well as Jhola Aandolan against plastic carry bags use, Global green mission, Operation water reservoir, stop global warming & climate changes, reduce pollution with dedication to save environmental and spiritual values.

Western Europe

and constituencies that value conservation and ecotourism above local subsistence. This has led to controversy among scientists and residents. Local herders see wild animals as unregulated public property subsidised by the work of the local people. Farmers complain that their fields are invaded on a daily basis by animals they cannot kill because of their protected status. Ranchers, under extremely strict sanitation regulations.

Eastern Europe

Ussr

In USSR the regime of Joseph Stalin (1924–53) concentrated on large-scale industrialisation, and has earned a historical reputation for paying little heed to the human and environmental costs of such rapid transformations. One exception was the logging industry, in which conflicting needs and ideologies enabled the preservation of significant tracts of forest.

Latvia

Galbreath and Auers (2009) examines history of environmental politics in Latvia, especially the formation of the Latvian Green Party, known as Zala Partija as part of the Green/Farmers' Union or Zalo un Zemnieku Savieniba (ZZP). The depoliticization of environmentalism emerged from the nationalist, corporate, and environmental elements of the ZZP. The three aspects of the environmentalism here are represented by the colours green, brown, and black. Green represents: ecological preservation and reversal of industrial side effects. Brown stands for civic and ethnic nationalism. Black represents the oil and gas pipeline industry and its influence on the 'Green' agenda in Latvia.

United States

Progressive Era

Both Conservationism and Environmentalism appeared in political debates during the Progressive Era in the early 20th century. There were three main positions. The laissez-faire position held that owners of private property—including lumber and mining companies, should be allowed to do anything they wished for their property.

The Conservationists, led by President Theodore Roosevelt and his close ally Gifford Pinchot, said that the laissez-faire approach was too wasteful and inefficient. In any case, they noted, most of the natural resources in the western states were already owned by the federal government. The best course of action, they argued, was a long-term plan devised by national experts to maximise the long-term economic benefits of natural resources.

Environmentalism was the third position, led by John Muir (1838–1914). Muir's passion for nature made him the most influential American environmentalist. Muir preached that nature was sacred and humans are intruders who should look but not develop. He founded the Sierra Club and remains an icon of the environmentalist movement. He was primarily responsible for defining the environmentalist position, in the debate between Conservation and environmentalism.

Environmentalism preached that nature was almost sacred, and that man was an intruder. It allowed for limited tourism (such as hiking), but opposed automobiles in national parks. It strenuously opposed timber cutting on most public lands, and vehemently denounced the dams that Roosevelt supported for water supplies, electricity and flood control. Especially controversial was the Hetch Hetchy dam in Yosemite National Park, which Roosevelt approved, and which supplies the water supply of San Francisco.

Theodore Roosevelt

Roosevelt put conservationist issue high on the national agenda. He worked with all the major figures of the movement, especially his chief advisor on the matter, Gifford Pinchot. Roosevelt was deeply committed to conserving natural resources, and is considered to be the nation's first conservation President. He encouraged the Newlands Reclamation Act of 1902 to promote federal construction of dams to irrigate small farms and placed 230 million acres (360,000 mi^2 or 930,000 km^2) under federal protection. Roosevelt set aside more Federal land for national parks and nature preserves than all of his predecessors combined. Roosevelt established the United States Forest Service, signed into law the creation of five National Parks, and signed the year 1906 Antiquities Act, under which he proclaimed 18 new U.S. National Monuments. He also established the first 51 Bird Reserves, four Game Preserves, and 150 National Forests, including Shoshone National Forest, the nation's first. The area of the United States that he placed under public protection totals approximately 230,000,000 acres (930,000 km^2).

Gifford Pinchot had been appointed by McKinley as chief of Division of Forestry in the Department of Agriculture. In 1905, his department gained control of the national forest reserves. Pinchot promoted private use (for a fee) under federal supervision. In 1907, Roosevelt designated 16 million acres (65,000 km^2) of new national forests just minutes before a deadline.

In May 1908, Roosevelt sponsored the Conference of Governors held in the White House, with a focus on natural resources and their most efficient use. Roosevelt delivered the opening address: "Conservation as a National Duty.".

In 1903 Roosevelt toured the Yosemite Valley with John Muir, who had a very different view of conservation, and tried to minimise commercial use of water resources and forests. Working through the Sierra Club he founded, Muir succeeded in 1905 in having Congress transfer the Mariposa Grove and Yosemite Valley to the Federal Government. While Muir wanted nature preserved for the sake of pure beauty, Roosevelt subscribed to Pinchot's formulation, "to make the forest produce the largest amount of whatever crop or service will be most useful, and keep on producing it for generation after generation of men and trees."

1930s

In the 1930s, the dominant view was the conservationism of Theodore Roosevelt, endorsed by Democrat Franklin D. Roosevelt, that led to the building of many large-scale dams and water projects, as well as the expansion of the National Forest System to buy out submarginal farms.

Since 1970

Environmental issues reemerged on the national agenda in 1970, with Republican Richard Nixon playing a major role, especially with his creation of the Environmental Protection Agency. The debates over the public lands and environmental politics played a supporting role in the decline of liberalism and the rise of modern conservatism. Although Americans consistently rank environmental issues as "important", polling data indicates that in the voting booth voters rank the environmental issues low relative to other political concerns.

The growth of the Republican party's political power in the inland West (apart from the Pacific coast) was facilitated by the rise of popular opposition to public lands reform. Successful Democrats in the inland West and Alaska typically take more conservative positions on environmental issues than Democrats from the Coastal states. Taking the conservationist position, conservatives drew on new organisational networks of think tanks, industry groups, and citizen-oriented organisations, and they began to deploy new strategies that affirmed the rights of individuals to their property, to hunt and recreate, and to pursue happiness unencumbered by the federal government.

Areas of Concern

Deforestation and overpopulation are issues affecting all regions of the world. The consequent destruction of wildlife habitat has prompted the creation of conservation groups in other countries, some founded by local hunters who have witnessed declining wildlife populations first hand. Also, it was highly important for the conservation movement to solve problems of living conditions in the cities and the overpopulation of such places.

Boreal Forest and the Arctic

The idea of incentive conservation is a modern one but its practice has clearly defended some of the sub Arctic wildernesses and the wildlife in those regions for thousands of years, especially by indigenous peoples such as the Evenk, Yakut, Sami, Inuit and Cree. The fur trade and hunting by these peoples have preserved these regions for thousands of years. Ironically, the pressure now upon them comes from non-renewable resources such as oil, sometimes to make synthetic clothing which is advocated as a humane substitute for fur. Similarly, in the case of the beaver, hunting and fur trade were thought to bring about the animal's demise, when in fact they were an integral part of its conservation. For many years children's books stated and still do, that the decline in the beaver population was due to the fur trade. In reality however, the decline in beaver numbers was because of habitat destruction and deforestation, as well as its continued persecution as a pest (it causes flooding). In Cree lands however, where the population valued the animal for meat and fur, it continued to thrive. The Inuit defend their relationship with the seal in response to outside critics.

Latin America (Bolivia)

The Izoceño-Guaraní of Santa Cruz Department, Bolivia is a tribe of hunters who were influential in establishing the Capitania del Alto y Bajo Isoso (CABI). CABI promotes economic growth and survival of the Izoceno people while discouraging the rapid destruction of habitat within Bolivia's Gran Chaco. They are responsible for the creation of the 34,000 square kilometre Kaa-Iya del Gran Chaco National Park and Integrated Management Area (KINP). The KINP protects the most biodiverse portion of the Gran Chaco, an ecoregion shared with Argentina, Paraguay and Brazil. In 1996, the Wildlife Conservation Society joined forces with CABI to institute wildlife and hunting monitoring programs in 23 Izoceño communities. The partnership combines traditional beliefs and local knowledge with the political and administrative tools needed to effectively manage habitats. The programs rely solely on voluntary participation by local hunters who perform self-monitoring techniques and keep records of their hunts. The information obtained by the hunters participating in the program has provided CABI with important data required to make educated decisions about the use of the land. Hunters have been willing participants in this program because of pride in their traditional activities, encouragement by their communities and expectations of benefits to the area.

Africa (Botswana)

In order to discourage illegal South African hunting parties and ensure future local use and sustainability, indigenous hunters in Botswana began lobbying for and implementing conservation practices in the 1960s. The Fauna Preservation Society of Ngamiland (FPS) was formed in 1962 by the husband and wife team: Robert Kay and June Kay, environmentalists working in conjunction with the Batawana tribes to preserve wildlife habitat. The FPS promotes habitat conservation and provides local education for preservation of wildlife.

Conservation initiatives were met with strong opposition from the Botswana government because of the monies tied to big-game hunting. In 1963, BaTawanga Chiefs and tribal hunter/adventurers in conjunction with the FPS founded Moremi National Park and Wildlife Refuge, the first area to be set aside by tribal people rather than governmental forces. Moremi National Park is home to a variety of wildlife, including lions, giraffes, elephants, buffalo, zebra, cheetahs and antelope, and covers an area of 3,000 square kilometres. Most of the groups involved with establishing this protected land were involved with hunting and were motivated by their personal observations of declining wildlife and habitat.

Conservation Biology

Conservation biology is the scientific study of the nature and status of Earth's biodiversity with the aim of protecting species, their habitats, and ecosystems from excessive rates of extinction. It is an interdisciplinary subject drawing on sciences, economics, and the practice of natural resource management.

History of Term

The term *conservation biology* was introduced as the title of a conference held at the University of California, San Diego in La Jolla, California in 1978 organised by biologists Bruce Wilcox and Michael E. Soulé. The meeting was prompted by the concern among scientists over tropical deforestation, disappearing species, eroding genetic diversity within species. The conference and proceedings that resulted sought to bridge a gap existing at the time between theory in ecology and population biology on the one hand and conservation policy and practice on the other. Conservation biology and the concept of biological diversity (biodiversity) emerged together, helping crystallize the modern era of conservation science and policy.

Description

The rapid decline of established biological systems around the world means that conservation biology is often referred to as a "Discipline with a deadline". Conservation biology is tied closely to ecology in researching the dispersal, migration, demographics, effective population size, inbreeding depression, and minimum population viability of rare or endangered species. To better understand the restoration ecology of native plant and animal communities, the conservation biologist closely studies both their polytypic and monotypic habitats that are affected by a wide range of benign and hostile factors. Conservation biology is concerned with phenomena that affect the maintenance, loss, and restoration of biodiversity and the science of sustaining evolutionary processes that engender genetic, population, species, and ecosystem diversity. The concern stems from estimates suggesting that up to 50% of all species on the planet will disappear within the next 50 years, which has contributed to poverty, starvation, and will reset the course of evolution on this planet.

Conservation biologists research and educate on the trends and process of biodiversity loss, species extinctions, and the negative effect these are having on our capabilities to sustain the well-being of human society. Conservation biologists work in the field and office, in government, universities, non-profit organisations and industry. They are funded to research, monitor, and catalogue every angle of the earth and its relation to society. The topics are diverse, because this is an interdisciplinary network with professional alliances in the biological as well as social sciences. Those dedicated to the cause and profession advocate for a global response to the current biodiversity crisis based on morals, ethics, and scientific reason. Organisations and citizens are responding to the biodiversity crisis through conservation action plans that direct research, monitoring, and education programs that engage concerns at local through global scales.

Context and Trends

Conservation biologists study trends and process from the paleontological past to the ecological present as they gain an understanding of the context related to species extinction. It is generally accepted that there have been five major global mass extinctions that register in Earth's history. These include: the Ordovician (440 mya), Devonian (370 mya), Permian–Triassic (245 mya), Triassic–Jurassic (200 mya), and Cretaceous (65 mya) extinction spasms. Within the last 10,000 years, human influence over the Earth's ecosystems has been so extensive that scientists have difficulty estimating the number of species lost; that is to say the rates of deforestation, reef destruction, wetland draining and other human acts are proceeding much faster than human assessment of species. The latest *Living Planet Report* by the World Wide Fund for Nature estimates that we have exceeded the bio-regenerative capacity of the planet, requiring 1.5 Earths to support the demands placed on our natural resources.

Sixth Extinction

Conservation biologists are dealing with and have published evidence from all corners of the planet indicating that humanity may be causing the sixth and greatest planetary extinction event. It has been suggested that we are living in an era of unprecedented numbers of species extinctions, also known as the Holocene extinction event. The global extinction rate may be approximately 100,000 times higher than the natural background extinction rate. It is estimated that two-thirds of all mammal genera and one-half of all mammal species weighing at least 44 kilograms (97 lb) have gone extinct in the last 50,000 years. It is speculated that this sixth extinction period is unique because it would be the first major extinction to be caused by another biotic agent over the course of the Earth's 4 billion year history. The Global Amphibian Assessment reports that amphibians are declining on a global scale faster than any other vertebrate group, with over 32% of all surviving species being threatened with extinction. The surviving populations are in continual decline in 43% of those that are threatened. Since the mid-1980s the actual rates of extinction have exceeded 211 times rates measured from the fossil record. However, "The current amphibian extinction rate may range from 25,039 to 45,474 times the background extinction rate for amphibians." The global extinction trend occurs in every major vertebrate group that is being monitored. For example, 23% of all mammals and 12% of all birds are Red Listed by the International Union for Conservation of Nature (IUCN), meaning they too are threatened with extinction.

Status of Oceans and Reefs

Global assessments of coral reefs of the world continue to report drastic and rapid rates of decline. By 2000, 27% of the world's coral reef ecosystems had effectively collapsed. The largest period of decline occurred in a dramatic "bleaching" event in 1998, where approximately 16% of all the coral reefs in the world disappeared in less than a year. *Coral bleaching* is caused by a mixture of environmental stresses, including increases in ocean temperatures and acidity, causing both the release of symbiotic algae and death of corals. Decline and extinction risk in coral reef biodiversity has risen dramatically in the past ten years. The loss of coral reefs, which are predicted to go extinct in the next century, will have huge economic impacts, threatens the balance of global biodiversity, and endangers food security for hundreds of millions of people. Conservation biology plays an important role in international agreements covering the world's oceans (and other issues pertaining to biodiversity, e.g.).

The oceans are threatened by acidification due to an increase in CO_2 levels. This is a most serious threat to societies relying heavily upon oceanic natural resources. A concern is that the majority of all marine species will not be able to evolve or acclimate in response to the changes in the ocean chemistry. The prospects of averting mass extinction seems unlikely when "[...] 90% of all of the large (average approximately e"50 kg), open ocean tuna, billfishes, and sharks in the ocean" are reportedly gone. Given the scientific review of current trends, the ocean is predicted to have few surviving multicellular organisms with only microbes left to dominate marine ecosystems.

Groups Other Than Vertebrates

Serious concerns also being raised about taxonomic groups that do not receive the same degree of social attention or attract funds as the vertebrates. These include fungal (including lichen-forming species), invertebrate (particularlyinsect) and plant communities where the vast majority of biodiversity is represented. Conservation of fungi and conservation of insects, in particular, are both of pivotal importance for conservation biology. As mycorrhizal symbionts, and as decomposers and recyclers, fungi are essential for sustainability of forests. The value of insects in the biosphere is enormous because they outnumber all other living groups in measure of species richness. The greatest bulk of biomass on land is found in plants, which is sustained by insect relations. This great ecological value of insects is countered by a society that oftentimes reacts negatively toward these aesthetically 'unpleasant' creatures.

One area of concern in the insect world that has caught the public eye is the mysterious case of missing honey bees (*Apis mellifera*). Honey bees provide an indispensable ecological services through their acts of pollination supporting a huge variety of agriculture crops. The sudden disappearance of bees leaving empty hives or colony collapse disorder (CCD) is not uncommon. However, in 16-month period from 2006 through 2007, 29% of 577 beekeepers across the United States reported CCD losses in up to 76% of their colonies. This sudden demographic loss in bee numbers is placing a strain on the agricultural sector. The cause behind the massive declines is puzzling scientists. Pests, pesticides, and global warming are all being considered as possible causes.

Another highlight that links conservation biology to insects, forests, and climate change is the mountain pine beetle (*Dendroctonus ponderosae*) epidemic of British Columbia, Canada, which has infested 470,000 km^2 (180,000 sq mi) of forested land since 1999. An action plan has been prepared by the Government of British Columbia to address this problem.

This impact [*pine beetle epidemic*] converted the forest from a small net carbon sink to a large net carbon source both during and immediately after the outbreak. In the worst year, the impacts resulting from the beetle outbreak in British Columbia were equivalent to 75% of the average annual direct forest fire emissions from all of Canada during 1959–1999. —Kurz *et al.*

Conservation Biology of Parasites

A large proportion of parasite species are threatened by extinction. A few of them are being eradicated as pests of humans or domestic animals, however, most of them are harmless. Threats include the decline or fragmentation of host populations, or the extinction of host species.

Threats to Biodiversity

Many of the threats to biodiversity, including disease and climate change, are reaching inside borders of protected areas, leaving them 'not-so protected' (e.g. Yellowstone National Park). Climate change, for example, is often cited as a serious threat in this regard, because there is a feedback loop between species extinction and the release of carbon dioxide into the atmosphere. Ecosystems store and cycle large amounts of carbon which regulates global conditions. The effects of global warming adds a catastrophic threat toward a mass extinction of global biological diversity. The extinction threat is estimated to range from 15 to 37 percent of all species by 2050, or 50 percent of all species over the next 50 years.

Some of the most significant and insidious threats to biodiversity and ecosystem processes include climate change, mass agriculture, deforestation, overgrazing, slash-and-burn agriculture, urban development, wildlife trade, light pollution and pesticide use. Habitat fragmentation poses one of the more difficult challenges, because the global network of protected areas only covers 11.5% of the Earth's surface. A significant consequence of fragmentation and lack of linked protected areas is the reduction of animal migration on a global scale. Considering that billions of tonnes of biomass are responsible for nutrient cycling across the earth, the reduction of migration is a serious matter for conservation biology.

Human activities are associated directly or indirectly with nearly every aspect of the current extinction spasm.

Wake and Vredenburg

These figures do not imply, however, that human activities must necessarily cause irreparable harm to the biosphere. With conservation management and planning for biodiversity at all levels, from genes to ecosystems, there are examples where humans mutually coexist in a sustainable way with nature. However, it may be too late for human intervention to reverse the current mass extinction.

Concepts and Foundations

Measuring Extinction Rates

Extinction rates are measured in a variety of ways. Conservation biologists measure and apply statistical measures of fossil records, rates of habitat loss, and a multitude of other variables such as loss of biodiversity as a function of the rate of habitat loss and site occupancy to obtain such estimates. The Theory of Island Biogeography is possibly the most significant contribution toward the scientific understanding of both the process and how to measure the rate of species extinction. The current background extinction rate is estimated to be one species every few years. The measure of ongoing species loss is made more complex by the fact that most of the Earth's species have not been described or evaluated. Estimates vary greatly on how many species actually exist (estimated range: 3,600,000-111,700,000) to how many have received a species binomial (estimated range: 1.5-8 million). Less than 1% of all species that have been described have been studied beyond simply noting its existence. From these figures, the IUCN reports that 23% of vertebrates, 5% of invertebrates and 70% of plants that have been evaluated are designated as endangered or threatened.

Systematic Conservation Planning

Systematic conservation planning is an effective way to seek and identify efficient and effective types of reserve design to capture or sustain the highest priority biodiversity values and to work with communities in support of local ecosystems. Margules and Pressey identify six interlinked stages in the systematic planning approach:

1. Compile data on the biodiversity of the planning region
2. Identify conservation goals for the planning region
3. Review existing conservation areas
4. Select additional conservation areas
5. Implement conservation actions
6. Maintain the required values of conservation areas.

Conservation biologists regularly prepare detailed conservation plans for grant proposals or to effectively coordinate their plan of action and to identify best management practices (e.g.). Systematic strategies generally employ the services of Geographic Information Systems to assist in the decision making process.

Conservation Biology as a Profession

The Society for Conservation Biology is a global community of conservation professionals dedicated to advancing the science and practice of conserving biodiversity. Conservation biology as a discipline reaches beyond biology, into subjects such as philosophy, law, economics, humanities, arts, anthropology, and education. Within biology, conservation genetics and evolution are immense fields unto themselves, but these disciplines are of prime importance to the practice and profession of conservation biology.

[...] there are advocates and there are sloppy or dishonest scientists, and these groups differ.

Chan: Is conservation biology an objective science when biologists advocate for an inherent value in nature? Do conservationists introduce bias when they support policies using qualitative description, such as habitat *degradation*, or *healthy* ecosystems? As all scientists hold values, so do conservation biologists. Conservation biologists advocate for reasoned and sensible management of natural resources and do so with a disclosed combination of science, reason, logic, and values in their conservation management plans. This sort of advocacy is similar to the medical profession advocating for healthy lifestyle options, both are beneficial to human well-being yet remain scientific in their approach. Many conservation biologists, in addition to having a Bachelors of Science (or extensive natural experience) often receive professional accreditation during their career (e.g.).

There is a movement in conservation biology suggesting a new form of leadership is needed to mobilise conservation biology into a more effective discipline that is able to communicate the full scope of the problem to society at large. The movement proposes an adaptive leadership approach that parallels an adaptive management approach. The concept is based on a new philosophy or leadership theory steering away from historical notions of power, authority, and dominance. Adaptive conservation leadership is reflective and more equitable as it applies to any member of society who can mobilise others toward meaningful change using communication techniques that are inspiring, purposeful, and collegial. Adaptive conservation leadership and mentoring programs are being implemented by conservation biologists through organisations such as the Aldo Leopold Leadership Program

Approaches

Conservation may be classified as either in-situ conservation, which is protecting an endangered species in its natural habitat, or ex-situ conservation, which occurs outside the natural habitat. In-situ conservation involves protecting or cleaning up the habitat itself which may include a great deal of environmental preservation, or by defending the species from predators. Ex-situ conservation may be used on some or all of the population, when in-situ conservation is too difficult, or impossible.

Also, non-interference may be used, which is termed a preservationist method. Preservationists advocate for giving areas of nature and species a protected existence that halts interference from the humans. In this regard, conservationists differ from preservationists in the social dimension, as conservation biology engages society and seeks equitable solutions for both society and ecosystems.

Some preservationists emphasize the potential of biodiversity in a world without humans:

> "Animals have not yet invaded 2/3 of Earth's habitats, and it could be that without human influence the diversity of tetrapods will continue to increase in an exponential fashion." —Sahney et al.

Ethics and Values

Conservation biologists are interdisciplinary researchers that practice ethics in the biological and social sciences. Chan states that conservationists must advocate for biodiversity and can do so in a scientifically ethical manner by not promoting simultaneous advocacy against other competing values. A conservationist researches biodiversity and reasons through a Resource Conservation Ethic, which identify what measures will deliver "the greatest good for the greatest number of people for the longest time."

Some conservation biologists argue that nature has an intrinsic value that is independent of anthropocentric usefulness or utilitarianism. Intrinsic value advocates that a gene, or species, be valued because they have a utility for the ecosystems they sustain. Aldo Leopold was a classical thinker and writer on such conservation ethics whose philosophy, ethics and writings are still valued and revisited by modern conservation biologists. His writing is oftentimes required reading for those in the profession.

Conservation Priorities

The International Union for the Conservation of Nature (IUCN) has organised a global assortment of scientists and research stations across the planet to monitor the changing state of nature in an effort to tackle the extinction crisis. The IUCN provides annual updates on the status of species conservation through its Red List. The IUCN Red

List serves as an international conservation tool to identify those species most in need of conservation attention and by providing a global index on the status of biodiversity. More than the dramatic rates of species loss, however, conservation scientists note that the sixth mass extinction is a biodiversity crisis requiring far more action than a priority focus on rare, endemic or endangered species. Concerns for biodiversity loss covers a broader conservation mandate that looks at ecological processes, such as migration, and a holistic examination of biodiversity at levels beyond the species, including genetic, population and ecosystem diversity. Extensive, systematic, and rapid rates of biodiversity loss threatens the sustained well-being of humanity by limiting supply of ecosystem services that are otherwise regenerated by the complex and evolving holistic network of genetic and ecosystem diversity. While the conservation status of species is employed extensively in conservation management, some scientists highlight that it is the common species that are the primary source of exploitation and habitat alteration by humanity. Moreover, common species are often undervalued despite their role as the primary source of ecosystem services.

While most in the community of conservation science "stress the importance" of sustaining biodiversity, there is debate on how to prioritize genes, species, or ecosystems, which are all components of biodiversity (e.g. Bowen, 1999). While the predominant approach to date has been to focus efforts on endangered species by conserving *biodiversity hotspots*, some scientists (e.g) and conservation organisations, such as the Nature Conservancy, argue that it is more cost effective, logical, and socially relevant to invest in *biodiversity coldspots*. The costs of discovering, naming, and mapping out the distribution every species, they argue, is an ill advised conservation venture. They reason it is better to understand the significance of the ecological roles of species.

Biodiversity hotspots and coldspots are a way of recognising that the spatial concentration of genes, species, and ecosystems is not uniformly distributed on the Earth's surface. For example, "[...] 44% of all species of vascular plants and 35% of all species in four vertebrate groups are confined to 25 hotspots comprising only 1.4% of the land surface of the Earth."

Those arguing in favour of setting priorities for coldspots point out that there are other measures to consider beyond biodiversity. They point out that emphasizing hotspots downplays the importance of the social and ecological connections to vast areas of the Earth's ecosystems where biomass, not biodiversity, reigns supreme. It is estimated that 36% of the Earth's surface, encompassing 38.9% of the worlds vertebrates, lacks the endemic species to qualify as biodiversity hotspot. Moreover, measures show that maximising protections for biodiversity does not capture ecosystem services any better than targeting randomly chosen regions. Population level biodiversity (i.e. coldspots) are disappearing at a rate that is ten times that at the species level. The level of importance in addressing biomass versus endemism as a concern for conservation biology is highlighted in literature measuring the level of threat to global ecosystem carbon stocks that do not necessarily reside in areas of endemism. A hotspot priority approach would not invest so heavily in places such as steppes, the Serengeti, the Arctic, or taiga. These areas contribute a great abundance of population (not species) level biodiversity and ecosystem services, including cultural value and planetary nutrient cycling.

Those in favour of the hotspot approach point out that species are irreplaceable components of the global ecosystem, they are concentrated in places that are most threatened, and should therefore receive maximal strategic protections. The IUCN Red List categories, which appear on Wikipedia species articles, is an example of the hotspot conservation approach in action; species that are not rare or endemic are listed the least concern and their wikipedia articles tend to be ranked low on the importance scale. This is a hotspot approach because the priority is set to target species level concerns over population level or biomass. Species richness and genetic biodiversity contributes to and engenders ecosystem stability, ecosystem processes, evolutionary adaptability, and biomass. Both sides agree, however, that conserving biodiversity is necessary to reduce the extinction rate and identify an inherent value in nature; the debate hinges on how to prioritize limited conservation resources in the most cost effective way.

Economic Values and Natural Capital

Conservation biologists have started to collaborate with leading global economists to determine how to measure the wealth and services of nature and to make these values apparent in global market transactions. This system of accounting is called *natural capital* and would, for example, register the value of an ecosystem before it is cleared to make way for development. The WWF publishes its *Living Planet Report* and provides a global index of biodiversity by monitoring approximately 5,000 populations in 1,686 species of vertebrate (mammals, birds, fish, reptiles, and amphibians) and report on the trends in much the same way that the stock market is tracked.

Tadrart Acacus desert in western Libya, part of the Sahara.

This method of measuring the global economic benefit of nature has been endorsed by the G8+5 leaders and the European Commission. Nature sustains many ecosystem services that benefit humanity. Many of the earths ecosystem services are public goods without a market and therefore no price or value. When the *stock market* registers a financial crisis, traders on Wall Street are not in the business of trading stocks for much of the planet's living natural capital stored in ecosystems. There is no natural stock market with investment portfolios into sea horses, amphibians, insects, and other creatures that provide a sustainable supply of ecosystem services that are valuable to society. The ecological footprint of society has exceeded the bio-regenerative capacity limits of the planet's ecosystems by about 30 percent, which is the same percentage of vertebrate populations that have registered decline from 1970 through 2005.

The ecological credit crunch is a global challenge. The Living Planet Report 2008 tells us that more than three quarters of the world's people live in nations that are ecological debtors – their national consumption has outstripped their country's biocapacity. Thus, most of us are propping up our current lifestyles, and our economic growth, by drawing (and increasingly overdrawing) upon the ecological capital of other parts of the world.

WWF Living Planet Report

The inherent natural economy plays an essential role in sustaining humanity, including the regulation of global atmospheric chemistry, pollinating crops, pest control, cycling soil nutrients, purifying our water supply, supplying medicines and health benefits, and unquantifiable quality of life improvements. There is a relationship, a correlation, between markets and natural capital, and social income inequity and biodiversity loss. This means that there are greater rates of biodiversity loss in places where the inequity of wealth is greatest.

Although a direct market comparison of natural capital is likely insufficient in terms of human value, one measure of ecosystem services suggests the contribution amounts to trillions of dollars yearly. For example, one segment of North American forests has been assigned an annual value of 250 billion dollars; as another example, honey-bee pollination is estimated to provide between 10 and 18 billion dollars of value yearly. The value of ecosystem services on one New Zealand island has been imputed to be as great as the GDP of that region. This planetary wealth is being lost at an incredible rate as the demands of human society is exceeding the bio-regenerative capacity of the Earth. While biodiversity and ecosystems are resilient, the danger of losing them is that humans cannot recreate many ecosystem functions through technological innovation.

Strategic Species Concepts

Keystone Species

Some species, called a *keystone species*, form a central supporting hub in the ecosystem. The loss of such a species results in a collapse in ecosystem function, as well as the loss of coexisting species. The importance of a keystone species was shown by the extinction of the Steller's Sea Cow (*Hydrodamalis gigas*) through its interaction with sea otters, sea urchins, and kelp. Kelp beds grow and form nurseries in shallow waters to shelter creatures that support the food chain. Sea urchins feed on kelp, while sea otters feed on sea urchins. With the rapid decline of sea otters due to overhunting, sea urchin populations grazed unrestricted on the kelp beds and the ecosystem collapsed. Left unchecked, the urchins destroyed the shallow water kelp communities that supported the Steller's Sea Cow's diet and hastened their demise. The sea otter is a keystone species because the coexistence of many ecological associates in the kelp beds relied upon otters for their survival.

Indicator Species

An *indicator species* has a narrow set of ecological requirements, therefore they become useful targets for observing the health of an ecosystem. Some animals, such as amphibians with their semi-permeable skin and linkages to wetlands, have an acute sensitivity to environmental harm and thus may serve as a *miner's canary*. Indicator species are monitored in an effort to capture environmental degradation through pollution or some other link to proximate human activities. Monitoring an indicator species is a measure to determine if there is a significant environmental impact that can serve to advise or modify practice, such as through different forestsilvi culture treatments and management scenarios, or to measure the degree of harm that a pesticide may impart on the health of an ecosystem.

Government regulators, consultants, or NGOs regularly monitor indicator species, however, there are limitations coupled with many practical considerations that must be followed for the approach to be effective. It is generally recommended that multiple indicators (genes, populations, species, communities, and landscape) be monitored for effective conservation measurement that prevents harm to the complex, and oftentimes unpredictable, response from ecosystem dynamics (Noss, 1997).

Umbrella and Flagship Species

An example of an *umbrella species* is the Monarch butterfly, because of its lengthy migrations and aesthetic value. The Monarch migrates across North America, covering multiple ecosystems and so requires a large area to exist. Any protections afforded to the Monarch butterfly will at the same time umbrella many other species and habitats. An umbrella species is often used as *flagship species*, which are species, such as the Giant Panda, the Blue Whale, the tiger, the mountain gorilla and the Monarch butterfly, that capture the public's attention and attract support for conservation measures.

History

The conservation of natural resources is the fundamental problem. Unless we solve that problem, it will avail us little to solve all others.

Theodore Roosevelt

Natural Resource Conservation

Efforts to conserve and protect *global* biodiversity are a recent phenomenon. Prior to the global conservation era, there was the coming of the age of conservation. Some historians have linked this with the 1916 National Parks Act, which included the 'use without impairment' clause, sought by John Muir. This eventually resulted in the removal of a proposal to build a dam in Dinosaur National Monument in 1959. Natural resource conservation, however, has a history that extends prior to the age of conservation. Resource ethics grew out of necessity through direct relations with nature. Regulation or communal restraint became necessary to prevent selfish motives from taking more than could be locally sustained, therefore compromising the long-term supply for the rest of the community. This social dilemma with respect to natural resource management is often called the "Tragedy of the Commons". From this principal, conservation biologists can trace communal resource based ethics throughout cultures as a solution to communal resource conflict. For example, the Alaskan Tlingit peoples and the Haida of the Pacific Northwest had resource boundaries, rules, and restrictions among clans with respect to the fishing of Sockeye Salmon. These rules were guided by clan elders who knew lifelong details of each river and stream they managed. There are numerous examples in history where cultures have followed rules, rituals, and organised practice with respect to communal natural resource management.

Conservation ethics are also found in early religious and philosophical writings. There are examples in the Tao, Shinto, Hindu, Islamic and Buddhist traditions. In Greek philosophy, Plato lamented about pasture land degradation: "What is left now is, so to say, the skeleton of a body wasted by disease; the rich, soft soil has been carried off and only the bare framework of the district left." In the bible, through Moses, God commanded to let the land rest from cultivation every seventh year. Before the 18th century, however, much of European culture considered it a pagan view to admire nature. Wilderness was denigrated while agricultural development was praised. However, as early as AD 680 a wildlife sanctuary was founded on the Farne Islands by St Cuthbert in response to his religious beliefs.

Early Naturalists

Natural history was a major preoccupation in the 18th century, with grand expeditions and the opening of popular public displays in Europe and North America. By 1900 there were 150 natural history museums in Germany, 250 in Great Britain, 250 in the United States, and 300 in France. Preservationist or conservationist sentiments are a development in the late 18th to early 20th century. The 19th century fascination with natural history engendered a fervor to be the first to collect rare specimens with the goal of doing so before they became extinct by other such collectors. Although his artistic works and romantic depiction of avian life inspired many bird enthusiasts and conservation organisations, the writings of John James Audubon, by modern standards, show insensitivity toward bird conservation as he shot and collected hundreds of specimens. Inspired by him, however, the first chapter of the Audubon Society started in 1905 for the purpose of protecting birds.

Coming of the Age of Conservation

The modern concept of ecosystem services can be found in the late 19th century. "The utility of Natural History or its applicability to promote the material wealth of the State cannot be doubted. It was a great mistake to suppose that the subjects of Zoology, Botany, and Geology did not involve much that affects our comfort, convenience, health and wealth." However, the article continues and discusses the dread of agricultural pests and the utility of understanding their natural history for the purpose of facilitating their destruction.

In the department of Woods and Forestry we should teach on the principals of conservation and teach on the lessons of economy rather than of waste in the natural resources of our country.

American Museum of Natural History, 1909

By the early 19th century biogeography was ignited through efforts of Alexander von Humboldt, Lyell and Darwin; their efforts, while important in relating species to their environments, were part of the naturalist tradition and fell short of conservation biology proper. Darwin, for example, hunted and shot birds and kept natural history cabinets in line with Victorian tradition.

Modern roots of conservation biology can be found in the late 19th century Enlightenment period particularly in England and Scotland. A number of thinkers, among them notably Lord Monboddo, described the importance of "preserving nature"; much of this early emphasis had its origins in Christian theology.

20th Century Conservation

In the 20th century, actions in the United Kingdom, United States, and Canada emphasized the protection of habitat areas pursuant to visions of such people as John Muir, Theodore Roosevelt, and Aldo Leopold. While the Canadian nor the United Kingdom governments did not pioneer the creation of National Parks as the United States did in the late 19th century, there were many farsighted civil servants who were dedicated to wildlife conservation and of notable mention. Some of these historical figures include Charles Gordon Hewitt and James Harkin.

The term *conservation* came into use in the late 19th century and referred to the management, mainly for economic reasons, of such natural resources as timber, fish, game, topsoil, pastureland, and minerals. In addition it referred to the preservation of forests (forestry), wildlife (wildlife refuge), parkland, wilderness, and watersheds. Western Europe was the source of much 19th century progress for conservation biology, particularly the British Empire with the Sea Birds Preservation Act 1869. However, the United States made contributions to this field starting with thinking of Thoreau and taking form with the Forest Act of 1891, John Muir's founding of the Sierra Club in 1892, the founding of the New York Zoological Society in 1895 and establishment of a series of national forests and preserves by Theodore Roosevelt from 1901 to 1909.

Not until the mid-20th century did efforts arise to target individual species for conservation, notably efforts in big cat conservation in South America led by the New York Zoological Society. In the early 20th century the New York Zoological Society was instrumental in developing concepts of establishing preserves for particular species and conducting the necessary conservation studies to determine the suitability of locations that are most appropriate as conservation priorities; the work of Henry Fairfield Osborn Jr., Carl E. Akeley, Archie Carr and Archie Carr III is notable in this era. Akeley for example, having led expeditions to the Virunga Mountains and observed the mountain gorilla in the wild, became convinced that the species and the area were conservation priorities. He was instrumental in persuading Albert I of Belgium to act in defence of the mountain gorilla and establish Albert National Park (since renamed Virunga National Park) in what is now Democratic Republic of Congo.

By the 1970s, led primarily by work in the United States under the Endangered Species Act along with the Species at Risk Act (SARA) of Canada, Biodiversity Action Plans developed in Australia, Sweden, the United Kingdom, hundreds of species specific protection plans ensued. Notably the United Nations acted to conserve sites of outstanding cultural or natural importance to the common heritage of mankind. The programme was adopted by the General Conference of UNESCO in 1972. As of 2006, a total of 830 sites are listed: 644 cultural, 162 natural. The first country to pursue aggressive biological conservation through national legislation was the United States, which passed back to back legislation in the Endangered Species Act (1966) and National Environmental Policy Act (1970), which together injected major funding and protection measures to large scale habitat protection and threatened species research. Other conservation developments, however, have taken hold throughout the world. India, for example, passed the Wildlife Protection Act of 1972.

In 1980 a significant development was the emergence of the urban conservation movement. A local organisation was established in Birmingham, UK, a development followed in rapid succession in cities across the UK, then overseas. Although perceived as a grassroots movement, its early development was driven by academic research into urban wildlife. Initially perceived as radical, the movement's view of conservation being inextricably linked with other human activity has now become mainstream in conservation thought. Considerable research effort is now directed at urban conservation biology. The Society for Conservation Biology originated in 1985.

By 1992 most of the countries of the world had become committed to the principles of conservation of biological diversity with the Convention on Biological Diversity; subsequently many countries began programmes of Biodiversity Action Plans to identify and conserve threatened species within their borders, as well as protect associated habitats. The late 1990s saw increasing professionalism in the sector, with the maturing of organisations such as the Institute of Ecology and Environmental Management and the Society for the Environment.

Since 2000 the concept of landscape scale conservation has risen to prominence, with less emphasis being given to single-species or even single-habitat focused actions. Instead an ecosystem approach is advocated by most mainstream conservationist, although concerns have been expressed by those working to protect some high-profile species.

Ecology has clarified the workings of the biosphere; i.e., the complex interrelationships among humans, other species, and the physical environment. The burgeoning human population and associated agriculture, industry, and the ensuing pollution, have demonstrated how easily ecological relationships can be disrupted.

"The last word in ignorance is the man who says of an animal or plant: "What good is it?" If the land mechanism as a whole is good, then every part is good, whether we understand it or not. If the biota, in the course of aeons, has built something we like but do not understand, then who but a fool would discard seemingly useless parts? To keep every cog and wheel is the first precaution of intelligent tinkering.

Conservation (Ethic)

Conservation is an ethic of resource use, allocation, and protection. Its primary focus is upon maintaining the health of the natural world: its, fisheries, habitats, and biological diversity. Secondary focus is on materials conservation and energy conservation, which are seen as important to protect the natural world. Those who follow the conservation ethic and, especially, those who advocate or work toward conservation goals are termed conservationists.

To conserve habitat in terrestrial ecoregions and stop deforestation is a goal widely shared by many groups with a wide variety of motivations.

To protect sea life from extinction due to overfishing is another commonly stated goal of conservation — ensuring that "some will be available for our children" to continue a way of life.

The consumer conservation ethic is sometimes expressed by the *four R's*: " Rethink, Reduce, Recycle, Repair" This social ethic primarily relates to local purchasing, moral purchasing, the sustained, and efficient use of renewable resources, the moderation of destructive use of finite resources, and the prevention of harm to common resources such as air and water quality, the natural functions of a living earth, and cultural values in a built environment.

The principal value underlying most expressions of the conservation ethic is that the natural world has intrinsic and intangible worth along with utilitarian value — a view carried forward by the scientific conservation movement and some of the older Romantic schools of ecology movement.

More Utilitarian schools of conservation seek a proper valuation of local and global impacts of human activity upon nature in their effect upon human well being, now and to our posterity. How such values are assessed and exchanged among people determines the social, political, and personal restraints and imperatives by which conservation is practiced. This is a view common in the modern environmental movement.

These movements have diverged but they have deep and common roots in the conservation movement.

In the United States of America, the year 1864 saw the publication of two books which laid the foundation for Romantic and Utilitarian conservation traditions in America. The posthumous publication of Henry David Thoreau's *Walden* established the grandeur of unspoiled nature as a citadel to nourish the spirit of man. From George Perkins Marsh a very different book, *Man and Nature*, later subtitled "The Earth as Modified by Human Action", catalogued his observations of man exhausting and altering the land from which his sustenance derives.

Terminology

The conservation of natural resources is the fundamental problem. Unless we solve that problem, it will avail us little to solve all others.

Theodore Roosevelt

In common usage, the term refers to the activity of systematically protecting natural resources such as forests, including biological diversity. Carl F. Jordan defines the term as:

> *biological conservation as being a philosophy of managing the environment in a manner that does not despoil, exhaust or extinguish.*

While this usage is not new, the idea of biological conservation has been applied to the principles of ecology, biogeography, anthropology, economy and sociology to maintain biodiversity.

The term "conservation" itself may cover the concepts such as cultural diversity, genetic diversity and the concept of movements environmental conservation, seedbank (preservation of seeds). These are often summarised as the priority to respect diversity, especially by Greens.

Much recent movement in conservation can be considered a resistance to commercialism and globalisation. Slow food is a consequence of rejecting these as moral priorities, and embracing a slower and more locally focused lifestyle.

Practice

Distinct trends exist regarding conservation development. While many countries' efforts to preserve species and their habitats have been government-led, those in the North Western Europe tended to arise out of the middle-class and aristocratic interest in natural history, expressed at the level of the individual and the national, regional or local learned society. Thus countries like Britain, the Netherlands, Germany, etc. had what we would today term NGOs — in the shape of the RSPB, National Trust and County Naturalists' Trusts (dating back to 1889, 1895 and 1912 respectively) Natuurmonumenten, Provincial conservation Trusts for each Dutch province, Vogelbescherming, etc. — a long time before there were National Parks and National Nature Reserves. This in part reflects the absence of wilderness areas in heavily cultivated Europe, as well as a longstanding interest in laissez-faire government in some countries, like the UK, leaving it as no coincidence that John Muir, the British-born founder of the National Park movement (and hence of government-sponsored conservation) did his sterling work in the USA, where he was the motor force behind the establishment of such NPs as Yosemite and Yellowstone. Nowadays, officially more than 10 percent of the world is legally protected in some way or the other, and in practice private fundraising is insufficient to pay for the effective management of so much land with protective status.

Protected areas in developing countries, where probably as many as 70-80 percent of the species of the world live, still enjoy very little effective management and protection. Some countries, such as Mexico, have non-profit civil organisations and land owners dedicated to protect vast private property, such is the case of Hacienda Chichen's Maya Jungle Reserve and Bird Refuge in Chichen Itza, Yucatán. The Adopt A Ranger Foundation has calculated that worldwide about 140,000 rangers are needed for the protected areas in developing and transition countries. There are no data on how many rangers are employed at the moment, but probably less than half the protected areas in developing and transition countries have any rangers at all and those that have them are at least 50% short This means that there would be a worldwide ranger deficit of 105,000 rangers in the developing and transition countries.

One of the world's foremost conservationists, Dr. Kenton Miller, stated about the importance of rangers: "The future of our ecosystem services and our heritage depends upon park rangers. With the rapidity at which the challenges to protected areas are both changing and increasing, there has never been more of a need for well prepared human capacity to manage. Park rangers are the backbone of park management. They are on the ground. They work on the front line with scientists, visitors, and members of local communities."

Adopt A Ranger, fears that the ranger deficit is the greatest single limiting factor in effectively conserving nature in 75% of the world. Currently, no conservation organisation or western country or international organisation addresses this problem. Adopt A Ranger has been incorporated to draw worldwide public attention to the most urgent problem that conservation is facing in developing and transition countries: protected areas without field staff. Very specifically, it will contribute to solving the problem by fund raising to finance rangers in the field. It will also help governments in developing and transition countries to assess realistic staffing needs and staffing strategies

Ecology

Ecology is the scientific study of the relations that living organisms have with respect to each other and their natural environment.

Variables of interest to ecologists include the composition, distribution, amount (biomass), number, and changing states of organisms within and among ecosystems. Ecosystems are hierarchical systems that are organised into a graded series of regularly interacting and semi-independent parts (e.g., species) that aggregate into higher orders of complex integrated wholes (e.g., communities). Ecosystems are sustained by the biodiversity within them. Biodiversity is the full-scale of life and its processes, including genes, species and ecosystems forming lineages that integrate into a complex and regenerative spatial arrangement of types, forms, and interactions. Ecosystems create biophysical feedback mechanisms between living (biotic) and nonliving (abiotic) components of the planet. These feedback loops regulate and sustain local communities, continental climate systems, and global biogeochemical cycles.

Ecology is a sub-discipline of biology, the study of life. The word "ecology" ("Ökologie") was coined in 1866 by the German scientist Ernst Haeckel (1834–1919). Ancient philosophers of Greece, including Hippocrates and Aristotle, were among the earliest to record notes and observations on the natural history of plants and animals. Modern ecology branched out of natural history and matured into a more rigorous science in the late 19th century. Charles Darwin's evolutionary treatise including the concept of adaptation, as it was introduced in 1859, is a pivotal cornerstone in modern ecological theory. Ecology is not synonymous with environment, environmentalism, natural history or environmental science. It is closely related to physiology, evolutionary biology, genetics and ethology. An understanding of how biodiversity affects ecological function is an important focus area in ecological studies. Ecologists seek to explain:

- Life processes and adaptations
- Distribution and abundance of organisms
- The movement of materials and energy through living communities
- The successional development of ecosystems, and
- The abundance and distribution of biodiversity in context of the environment.

Ecology is a human science as well. There are many practical applications of ecology in conservation biology, wetland management, natural resource management (agriculture, forestry, fisheries), city planning (urban ecology), community health, economics, basic and applied science and human social interaction (human ecology). Ecosystems sustain every life-supporting function on the planet, including climate regulation, water filtration, soil formation (pedogenesis), food, fibres, medicines, erosion control, and many other natural features of scientific, historical or spiritual value.

Integrative Levels, Scope, and Scale of Organisation

The scope of ecology covers a wide array of interacting levels of organisation spanning micro-level (e.g., cells) to planetary scale (e.g., ecosphere) phenomena. Ecosystems, for example, contain populations of individuals that aggregate into distinct ecological communities. It can take thousands of years for ecological processes to mature through and until the final successional stages of a forest. The area of an ecosystem can vary greatly from tiny to vast. A single tree is of little consequence to the classification of a forest ecosystem, but critically relevant to the smaller organisms living in and on it. Several generations of an aphid population can exist over the lifespan of a single leaf. Each of those aphids, in turn, support diverse bacterial communities. The nature of connections in ecological communities cannot be explained by knowing the details of each species in isolation, because the emergent pattern is neither revealed nor predicted until the ecosystem is studied as an integrated whole. Some ecological principles, however, do exhibit collective properties where the sum of the components explain the properties of the whole, such as birth rates of a population being equal to the sum of individual births over a designated time frame.

Hierarchical Ecology

The scale of ecological dynamics can operate like a closed island with respect to local site variables, such as aphids migrating on a tree, while at the same time remain open with regard to broader scale influences, such as atmosphere or climate. Hence, ecologists have devised means of hierarchically classifying ecosystemsby analysing data collected from finer scale units, such as vegetation associations, climate, and soil types, and integrate this information to identify larger emergent patterns of uniform organisation and processes that operate on local to regional, landscape, and chronological scales.

To structure the study of ecology into a manageable framework of understanding, the biological world is conceptually organised as a nested hierarchy of organisation, ranging in scale from genes, to cells, to tissues, to organs, to organisms, to species and up to the level of the biosphere. Together these hierarchical scales of life form a panarchy and they exhibit non-linear behaviours; "nonlinearity refers to the fact that effect and cause are disproportionate, so that small changes in critical variables, such as the numbers of nitrogen fixers, can lead to disproportionate, perhaps irreversible, changes in the system properties."

Biodiversity

Biodiversity (an abbreviation of biological diversity) describes the diversity of life from genes to ecosystems and spans every level of biological organisation. Biodiversity means different things to different people and there are many ways to index, measure, characterise, and represent its complex organisation. Biodiversity includes species diversity, ecosystem diversity, genetic diversity and the complex processes operating at and among these respective levels. Biodiversity plays an important role in ecological health as much as it does for human health. Preventing or prioritizing species extinctions is one way to preserve biodiversity, but populations, the genetic diversity within them and ecological processes, such as migration, are being threatened on global scales and disappearing rapidly as well. Conservation priorities and management techniques require different approaches and considerations to address the full ecological scope of biodiversity. Populations and species migration, for example, are more sensitive indicators of ecosystem services that sustain and contribute natural capital toward the well-being of humanity. An understanding of biodiversity has practical application for ecosystem-based conservation planners as they make ecologically responsible decisions in management recommendations to consultant firms, governments and industry.

Habitat

The habitat of a species describes the environment over which a species is known to occur and the type of community that is formed as a result. More specifically, "habitats can be defined as regions in environmental space that are composed of multiple dimensions, each representing a biotic or abiotic environmental variable; that is, any component or characteristic of the environment related directly (e.g. forage biomass and quality) or indirectly (e.g. elevation) to the use of a location by the animal." For example, the habitat might refer to an aquatic or terrestrial environment that can be further categorised as montane or alpine ecosystems. Habitat shifts provide important evidence of competition in nature where one population changes relative to the habitats that most other individuals of the species occupy. One population of a species of tropical lizards (*Tropidurus hispidus*), for example, has a flattened body relative to the main populations that live in open savanna. The population that lives in an isolated rock outcrop hides in crevasses where its flattened body may improve its performance. Habitat shifts also occur in the developmental life history of amphibians and many insects that transition from aquatic to terrestrial habitats. Biotope and habitat are sometimes used interchangeably, but the former applies to a communities environment, whereas the latter applies to a species' environment.

Niche

Termite mounds with varied heights of chimneys regulate gas exchange, temperature and other environmental parameters that are needed to sustain the internal physiology of the entire colony.

There are many definitions of the niche dating back to 1917, but G. Evelyn Hutchinson made conceptual advances in 1957 and introduced the most widely accepted definition: "which a species is able to persist and maintain stable population sizes." The ecological niche is a central concept in the ecology of organisms and is subdivided into the *fundamental* and the *realised* niche. The fundamental niche is the set of environmental conditions under which a species is able to persist. The realised niche is the set of environmental plus ecological conditions under which a species persists. The Hutchisonian niche is defined more technically as an "Euclidean hyperspace whose *dimensions* are defined as environmental variables and whose *size* is a function of the number of values that the environmental values may assume for which an organism has *positive fitness*." Biogeographical patterns and range distributions are explained or predicted through knowledge and understanding of a species traits and niche requirements. Species have functional traits that are uniquely adapted to the ecological niche. A trait is a measurable property, phenotype, or characteristic of an organism that influences its performance. Genes play an important role in the development and expression of traits. Resident species evolve traits that are fitted to their local environment. This tends to afford them a competitive advantage and discourages similarly adapted species from having an overlapping geographic range. The competitive exclusion principle suggests that two species cannot coexist indefinitely by living off the same limiting resource. When similarly adapted species are found to overlap geographically, closer inspection reveals subtle ecological differences in their habitat or dietary requirements. Some models and empirical studies, however, suggest that disturbances can stabilise the coevolution and shared niche occupancy of similar species inhabiting species rich communities. The habitat plus the niche is called the ecotope, which is defined as the full range of environmental and biological variables affecting an entire species.

Niche Construction

Organisms are subject to environmental pressures, but they are also modifiers of their habitats. The regulatory feedback between organisms and their environment can modify conditions from local (e.g., a beaver pond) to global scales (e.g., Gaia), over time and even after death, such as decaying logs or silica skeleton deposits from marine organisms. The process and concept of ecosystem engineering has also been called niche construction. Ecosystem engineers are defined as: "...organisms that directly or indirectly modulate the availability of resources to other species, by causing physical state changes in biotic or abiotic materials. In so doing they modify, maintain and create habitats."

The ecosystem engineering concept has stimulated a new appreciation for the degree of influence that organisms have on the ecosystem and evolutionary process. The terms niche construction are more often used in reference to the under appreciated feedback mechanism of natural selection imparting forces on the abiotic niche. An example of natural selection through ecosystem engineering occurs in the nests of social insects, including ants, bees, wasps, and termites. There is an emergent homeostasis or homeorhesis in the structure of the nest that regulates, maintains and defends the physiology of the entire colony. Termite mounds, for example, maintain a constant internal temperature through the design of air-conditioning chimneys.

The structure of the nests themselves are subject to the forces of natural selection. Moreover, the nest can survive over successive generations, which means that ancestors inherit both genetic material and a legacy niche that was constructed before their time.

Biome

Biomes are larger units of organisation that categorise regions of the Earth's ecosystems mainly according to the structure and composition of vegetation. Different researchers have applied different methods to define continental boundaries of biomes dominated by different functional types of vegetative communities that are limited in distribution by climate, precipitation, weather and other environmental variables. Examples of biome names include: tropical rainforest, temperate broadleaf and mixed forests, temperate deciduous forest, taiga, tundra, hot desert, and polar desert. Other researchers have recently started to categorise other types of biomes, such as the human and oceanic microbiomes. To a microbe, the human body is a habitat and a landscape. The microbiome has been largely discovered through advances in molecular genetics that have revealed a hidden richness of microbial diversity on the planet. The oceanic microbiome plays a significant role in the ecological biogeochemistry of the planet's oceans.

Biosphere

Ecological theory has been used to explain self-emergent regulatory phenomena at the planetary scale. The largest scale of ecological organisation is the biosphere: the total sum of ecosystems on the planet. Ecological relationships regulate the flux of energy, nutrients, and climate all the way up to the planetary scale. For example, the dynamic history of the planetary CO_2 and O_2 composition of the atmosphere has been largely determined by the biogenic flux of gases coming from respiration and photosynthesis, with levels fluctuating over time and in relation to the ecology and evolution of plants and animals.

When sub-component parts are organised into a whole there are oftentimes emergent properties that describe the nature of the system. The Gaia hypothesis is an example of holism applied in ecological theory. The ecology of the planet acts as a single regulatory or holistic unit called Gaia. The Gaia hypothesis states that there is an emergent feedback loop generated by the metabolism of living organisms that maintains the temperature of the Earth and atmospheric conditions within a narrow self-regulating range of tolerance.

Population Ecology

The population is the unit of analysis in population ecology. A population consists of individuals of the same species that live, interact and migrate through the same niche and habitat. A primary law of population ecology is the Malthusian growth model. This law states that:

"...a population will grow (or decline) exponentially as long as the environment experienced by all individuals in the population remains constant."

This Malthusian premise provides the basis for formulating predictive theories and tests that follow. Simplified population models usually start with four variables including death, birth, immigration, and emigration. Mathematical models are used to calculate changes in population demographics using a null model. A null model is used as a null hypothesis for statistical testing. The null hypothesis states that random processes create observed patterns. Alternatively the patterns differ significantly from the random model and require further explanation. Models can be mathematically complex where "...several competing hypotheses are simultaneously confronted with the data." An example of an introductory population model describes a closed population, such as on an island, where immigration and emigration does not take place. In these island models the rate of population change is described by:

$$\frac{dN}{dT} = B - D = bN - dN = (b - d)N = rN,$$

where N is the total number of individuals in the population, B is the number of births, D is the number of deaths, b and d are the per capita rates of birth and death respectively, and r is the per capita rate of population change. This formula can be read out as the rate of change in the population (dN/dT) is equal to births minus deaths ($B - D$).

Using these modelling techniques, Malthus' population principle of growth was later transformed into a model known as the logistic equation:

$$\frac{dN}{dT} = aN\left(1 - \frac{N}{K}\right),$$

where N is the number of individuals measured as biomass density, a is the maximum per-capita rate of change, and K is the carrying capacity of the population. The formula can be read as follows: the rate of change in the population (dN/dT) is equal to growth (aN) that is limited by carrying capacity ($1 - N/K$). The discipline of population ecology builds upon these introductory models to further understand demographic processes in real study populations and conduct statistical tests. The field of population ecology often uses data on life history and matrix algebra to develop projection matrices on fecundity and survivorship. This information is used for managing wildlife stocks and setting harvest quotas.

Metapopulations and Migration

Populations are also studied and modelled according to the metapopulation concept. The metapopulation concept was introduced in 1969: "as a population of populations which go extinct locally and recolonise." Metapopulation ecology is another statistical approach that is often used in conservation research. Metapopulation research simplifies the landscape into patches of varying levels of quality. Metapopulations are linked by the migratory behaviours of organisms. Animal migration is set apart from other kinds of movement because it involves the seasonal departure and return of individuals from one habitat to another.

Migration is also a population level phenomenon, such as the migration routes followed by plants as they occupied northern post-glacial environments.

Plant ecologists rely on pollen records that accumulate and stratify in wetlands to reconstruct the timing of plant migration and dispersal relative to historic and contemporary climates. These migration routes involved an expansion of the range as plant populations expanded from one area to another. There is a larger taxonomy of movement, such as commuting, foraging, territorial behaviour, stasis, and ranging. Dispersal is usually distinguished from migration because it involves the one way permanent movement of individuals from their birth population into another population.

In metapopulation terminology there are emigrants (individuals that leave a patch), immigrants (individuals that move into a patch) and sites are classed either as sources or sinks. A site is a generic term that refers to places where ecologists sample populations, such as ponds or defined sampling areas in a forest. Source patches are productive sites that generate a seasonal supply of juveniles that migrate to other patch locations. Sink patches are unproductive sites that only receive migrants and will go extinct unless rescued by an adjacent source patch or environmental conditions become more favourable. Metapopulation models examine patch dynamics over time to answer questions about spatial and demographic ecology. The ecology of metapopulations is a dynamic process of extinction and colonisation.

Small patches of lower quality (i.e., sinks) are maintained or rescued by a seasonal influx of new immigrants. A dynamic metapopulation structure evolves from year to year, where some patches are sinks in dry years and become sources when conditions are more favourable. Ecologists use a mixture of computer models and field studies to explain metapopulation structure.

Community Ecology

Interspecific interactions such as predation are a key aspect of community ecology.

Community ecology examines how interactions among species and their environment affect the abundance, distribution and diversity of species within communities.

Johnson & Stinchcomb

Community ecology is the study of the interactions among a collection of interdependent species that cohabitate the same geographic area. An example of a study in community ecology might measure primary production in a wetland in relation to decomposition and consumption rates. This requires an understanding of the community connections between plants (i.e., primary producers) and the decomposers (e.g., fungi and bacteria) or the analysis of predator-prey dynamics affecting amphibian biomass. Food webs and trophic levels are two widely employed conceptual models used to explain the linkages among species.

Ecosystem Ecology

A riparian forest in the White Mountains, New Hampshire (USA), an example of ecosystem ecology

The concept of the ecosystem was fully synthesized in 1935 to describe habitats within biomes that form an integrated whole and a dynamically responsive system having both physical and biological complexes. However, the underlying concept can be traced back to 1864 in the published work of George Perkins Marsh ("Man and Nature"). Within an ecosystem there are inseparable ties that link organisms to the physical and biological components of their environment to which they are adapted. Ecosystems are complex adaptive systems where the interaction of life processes form self-organising patterns across different scales of time and space. Terrestrial, freshwater, atmospheric, and marine ecosystems very broadly cover the major types. Differences stem from the nature of the unique physical environments that shapes the biodiversity within each. A more recent addition to ecosystem ecology are the novel technoecosystems of the anthropocene.

Food Webs

A food web is the archetypal ecological network. Plants capture and convert solar energy into the biomolecular bonds of simple sugars during photosynthesis. This food energy is transferred through a series of organisms starting with those that feed on plants and are themselves consumed. The simplified linear feeding pathways that move from a basal trophic species to a top consumer is called the food chain. The larger interlocking pattern of food chains in an ecological community creates a complex food web. Food webs are a type of concept map or a heuristic device that is used illustrate and study pathways of energy and material flows.

Food webs are often limited relative to the real world. Complete empirical measurements are generally restricted to a specific habitat, such as a cave or a pond. Principles gleaned from food web microcosm studies are used to extrapolate smaller dynamic concepts to larger systems. Feeding relations require extensive investigations into the gut of organisms, which can be very difficult to decipher, or (more recently) stable isotopes can be used to trace the flow of nutrient diets and energy through a food web. While food webs often give an incomplete measure of ecosystems, they are nonetheless a valuable tool in understanding community ecosystems.

Food-webs exhibit principals of ecological emergence through the nature of trophic entanglement, where some species have many weak feeding links (e.g., omnivores) while some are more specialised with fewer stronger feeding links (e.g., primary predators). Theoretical and empirical studies identify non-random emergent patterns of few strong and many weak linkages that serve to explain how ecological communities remain stable over time. Food-webs have compartments, where the many strong interactions create subgroups among some members in a community and the few weak interactions occur between these subgroups. These compartments increase the stability of food-webs. As plants grow, they accumulate carbohydrates and are eaten by grazing herbivores. Step by step lines or relations are drawn until a web of life is illustrated.

Trophic Levels

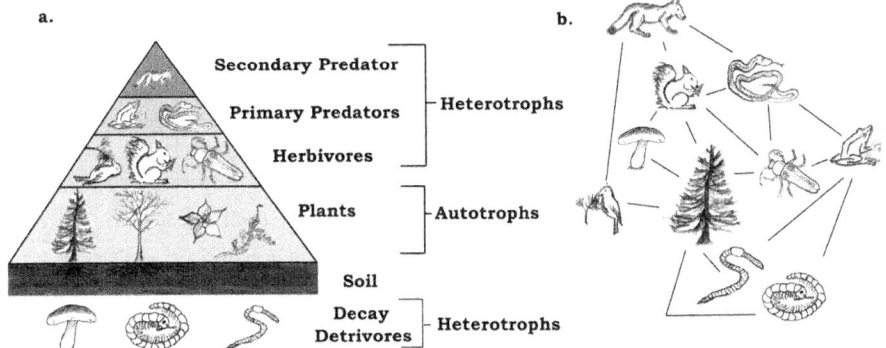

A trophic pyramid (a) and a food-web (b) illustrating ecological relationships among creatures that are typical of a northern Boreal terrestrial ecosystem. The trophic pyramid roughly represents the biomass (usually measured as total dry-weight) at each level. Plants generally have the greatest biomass. Names of trophic categories are shown to the right of the pyramid. Some ecosystems, such as many wetlands, do not organise as a strict pyramid, because aquatic plants are not as productive as long-lived terrestrial plants such as trees. Ecological trophic pyramids are typically one of three kinds: 1) pyramid of numbers, 2) pyramid of biomass, or 3) pyramid of energy.

The Greek root of the word *troph*, means food or feeding. Links in food-webs primarily connect feeding relations or trophism among species. Biodiversity within ecosystems can be organised into vertical and horizontal dimensions. The vertical dimension represents feeding relations that become further removed from the base of the food chain up toward top predators. A trophic level is defined as "a group of organisms acquiring a considerable majority of its energy from the adjacent level nearer the abiotic source." The horizontal dimension represents the abundance or biomass at each level. When the relative abundance or biomass of each functional feeding group is stacked into their respective trophic levels they naturally sort into a 'pyramid of numbers'.

Functional groups are broadly categorised as autotrophs (or primary producers), heterotrophs (or consumers), and detrivores (or decomposers). Autotrophs are organisms that can produce their own food (production is greater than respiration) and are usually plants or cyanobacteria that are capable of photosynthesis but can also be other organisms such as bacteria near ocean vents that are capable of chemosynthesis. Heterotrophs are organisms that must feed on others for nourishment and energy (respiration exceeds production). Heterotrophs can be further sub-divided into different functional groups, including: primary consumers (strict herbivores), secondary consumers (carnivorous predators that feed exclusively on herbivores) and tertiary consumers (predators that feed on a mix of herbivores and predators). Omnivores do not fit neatly into a functional category because they eat both plant and animal tissues. It has been suggested that omnivores have a greater functional influence as predators because relative to herbivores they are comparatively inefficient at grazing.

Trophic levels are part of the holistic or complex systems view of ecosystems. Each trophic level contains unrelated species that grouped together because they share common ecological functions. Grouping functionally similar species into a trophic system gives a macroscopic image of the larger functional design. While the notion of trophic levels provides insight into energy flow and top-down control within food webs, it is troubled by the prevalence of omnivory in real ecosystems. This has lead some ecologists to "reiterate that the notion that species clearly aggregate into discrete, homogeneous trophic levels is fiction." Nonetheless, recent studies have shown that real trophic levels do exist, but "above the herbivore trophic level, food webs are better characterised as a tangled web of omnivores."

Keystone Species

A keystone species is a species that is disproportionately connected to more species in the food-web. Keystone species have lower levels of biomass in the trophic pyramid relative to the importance of their role. The many connections that a keystone species holds means that it maintains the organisation and structure of entire communities. The loss of a keystone species results in a range of dramatic cascading effects that alters trophic dynamics, other food-web connections and can cause the extinction of other species in the community.

Sea otters (*Enhydra lutris*) are commonly cited as an example of a keystone species because they limit the density of sea urchins that feed on kelp. If sea otters are removed from the system, the urchins graze until the kelp beds disappear and this has a dramatic effect on community structure. Hunting of sea otters, for example, is thought to have indirectly led to the extinction of the Steller's Sea Cow (*Hydrodamalis gigas*). While the keystone species concept has been used extensively as a conservation tool, it has been criticized for being poorly defined from an operational stance. It is very difficult to experimentally determine in each different ecosystem what species may hold a keystone role. Furthermore, food-web theory suggests that keystone species may not be all that common. It is therefore unclear how generally the keystone species model can be applied.

Soils

Soil is the living top layer of mineral and organic dirt that covers the surface of the planet, it is the chief organising centre of most ecosystem functions, and it is of critical importance in agricultural science and ecology. The decomposition of dead organic matter, such as leaves falling on the forest floor, turns into soils containing minerals and nutrients that feed into plant production. The total sum of the planet's soil ecosystems is called the pedosphere where a very large proportion of the Earth's biodiversity sorts into other trophic levels. Invertebrates that feed and shred larger leaves, for example, create smaller bits for smaller organisms in the feeding chain. Collectively, these are the detrivores that regulate soil formation. Tree roots, fungi, bacteria, worms, ants, beetles, centipedes, spiders, mammals, birds, reptiles, amphibians and other less familiar creatures all work to create the trophic web of life in soil ecosystems. As organisms feed and migrate through soils they physically displace materials, which is an important ecological process called bioturbation. Bioturbation helps to aerate the soils, thus stimulating hetertrophic growth and production. Biomass of soil microorganisms are influenced by and feed back into the trophic dynamics of the exposed solar surface ecology. Paleoecological studies of soils places the origin for bioturbation to a time before the Cambrian period. Other events, such as the evolution of trees and amphibians moving into land in the Devonian period played a significant role in the development of the ecological trophism in soils.

Ecological Complexity

Complexity is easily understood as a large computational effort needed to piece together numerous interacting parts exceeding the iterative memory capacity of the human mind. Global patterns of biological diversity are complex. This biocomplexity stems from the interplay among ecological processes that operate and influence patterns at different scales that grade into each other, such as transitional areas or ecotones spanning landscapes. Complexity stems from the interplay among levels of biological organisation as energy and matter is integrated into larger units that superimpose onto the smaller parts. "What were wholes on one level become parts on a higher one." Small scale patterns do not necessarily explain large scale phenomena, otherwise captured in the expression (coined by Aristotle) 'the sum is greater than the parts'.

"Complexity in ecology is of at least six distinct types: spatial, temporal, structural, process, behavioural, and geometric." Out of these principles, ecologists have identified emergent and self-organising phenomena that operate at different environmental scales of influence, ranging from molecular to planetary, and these require different sets of scientific explanation at each integrative level. Ecological complexity relates to the dynamic resilience of ecosystems that transition to multiple shifting steady-states directed by random fluctuations of history. Long-term ecological studies provide important track records to better understand the complexity and resilience of ecosystems over longer temporal and broader spatial scales. The International Long Term Ecological Network manages and exchanges scientific information among research sites. The longest experiment in existence is the Park Grass Experiment that was initiated in 1856. Another example includes the Hubbard Brook study in operation since 1960.

Holism

The biological organisation of life self-organises into layers of emergent whole systems that function according to nonreducible properties called holism. This means that higher order patterns of a whole functional system, such as an ecosystem, cannot be predicted or understood by a simple summation of the parts. "New properties emerge because the components interact, not because the basic nature of the components is changed."

Ecological studies are necessarily holistic as opposed to reductionistic. Holism has three scientific meanings or uses that identify with: 1) the mechanistic complexity of ecosystems, 2) the practical description of patterns in quantitative reductionist terms where correlations may be identified but nothing is understood about the causal relations without reference to the whole system, which leads to 3) a metaphysical hierarchy whereby the causal relations of larger systems are understood without reference to the smaller parts. An example of the metaphysical aspect to holism is identified in the trend of increased exterior thickness in shells of different species. The reason for a thickness increase can be understood through reference to principals of natural selection via predation without need to reference or understand the biomolecular properties of the exterior shells.

Relation to Evolution

Ecology and evolution are considered sister disciplines of the life sciences. Natural selection, life history, development, adaptation, populations, and inheritance are examples of concepts that thread equally into ecological and evolutionary theory. Morphological, behavioural and/or genetic traits, for example, can be mapped onto evolutionary trees to study the historical development of a species in relation to their functions and roles in different ecological circumstances. In this framework, the analytical tools of ecologists and evolutionists overlap as they organise, classify and investigate life through common systematic principals, such as phylogenetics or the Linnaean system of taxonomy. The two disciplines often appear together, such as in the title of the journal *Trends in Ecology and Evolution*. There is no sharp boundary separating ecology from evolution and they differ more in their areas of applied focus. Both disciplines discover and explain emergent and unique properties and processes operating across different spatial or temporal scales of organisation. While the boundary between ecology and evolution is not always clear, it is understood that ecologists study the abiotic and biotic factors that influence the evolutionary process.

Behavioural Ecology

All organisms are motile to some extent. Even plants express complex behaviour, including memory and communication. Behavioural ecology is the study of ethology and its ecological and evolutionary implications. Ethology is the study of observable movement or behaviour in nature. This could include investigations of motile sperm of plants, mobile phytoplankton, zooplankton swimming toward the female egg, the cultivation of fungi by weevils, the mating dance of a salamander, or social gatherings of amoeba.

Adaptation is the central unifying concept in behavioural ecology. Behaviours can be recorded as traits and inherited in much the same way that eye and hair colour can. Behaviours evolve and become adapted to the ecosystem because they are subject to the forces of natural selection. Hence, behaviours can be adaptive, meaning that they evolve functional utilities that increases reproductive success for the individuals that inherit such traits. This is also the technical definition for fitness in biology, which is a measure of reproductive success over successive generations.

Predator-prey interactions are an introductory concept into food-web studies as well as behavioural ecology. Prey species can exhibit different kinds of behavioural adaptations to predators, such as avoid, flee or defend.

Many prey species are faced with multiple predators that differ in the degree of danger posed. To be adapted to their environment and face predatory threats, organisms must balance their energy budgets as they invest in different aspects of their life history, such as growth, feeding, mating, socializing, or modifying their habitat. Hypotheses posited in behavioural ecology are generally based on adaptive principals of conservation, optimisation or efficiency. For example,

"The threat-sensitive predator avoidance hypothesis predicts that prey should assess the degree of threat posed by different predators and match their behaviour according to current levels of risk."

"The optimal flight initiation distance occurs where expected postencounter fitness is maximised, which depends on the prey's initial fitness, benefits obtainable by not fleeing, energetic escape costs, and expected fitness loss due to predation risk."

Elaborate sexual displays and posturing are encountered in the behavioural ecology of animals. The birds of paradise, for example, display elaborate ornaments and song during courtship. These displays serve a dual purpose of signaling healthy or well-adapted individuals and desirable genes. The elaborate displays are driven by sexual selection as an advertisement of quality of traits among male suitors.

Symbiosis: Leafhoppers (*Eurymela fenestrata*) are protected by ants (*Iridomyrmex purpureus*) in a symbioticrelationship. The ants protect the leafhoppers from predators and in return the leafhoppers feeding on plants exude honeydew from their anus that provides energy and nutrients to tending ants.

Social Ecology

Social ecological behaviours are notable in the social insects, slime moulds, social spiders, human society, and naked mole rats where eusocialism has evolved. Social behaviours include reciprocally beneficial behaviours among kin and nest mates. Social behaviours evolve from kin and group selection. Kin selection explains altruism through genetic relationships, whereby an altruistic behaviour leading to death is rewarded by the survival of genetic copies distributed among surviving relatives. The social insects, including ants, bees and wasps are most famously studied for this type of relationship because the male drones are clones that share the same genetic make-up as every other male in the colony. In contrast, group selectionists find examples of altruism among non-genetic relatives and explain this through selection acting on the group, whereby it becomes selectively advantageous for groups if their members express altruistic behaviours to one another. Groups that are predominantly altruists beat groups that are predominantly selfish.

Coevolution

Ecological interactions can be divided into host and associate relationships. A host is any entity that harbours another that is called the associate. Host and associate relationships among species that are mutually or reciprocally beneficial are called mutualisms. If the host and associate are physically connected, the relationship is called symbiosis. Approximately 60% of all plants, for example, have a symbiotic relationship with arbuscular mycorrhizal fungi. Symbiotic plants and fungi exchange carbohydrates for mineral nutrients. Symbiosis differs from indirect mutualisms where the organisms live apart. For example, tropical rainforests regulate the Earth's atmosphere. Trees living in the equatorial regions of the planet supply oxygen into the atmosphere that sustains species living in distant polar regions of the planet. This relationship is called commensalism because many other host species receive the benefits of clean air at no cost or harm to the associate tree species supplying the oxygen. The host and associate relationship is called parasitism if one species benefits while the other suffers. Competition among species or among members of the same species is defined as reciprocal antagonism, such as grasses competing for growth space.

Popular ecological study systems for mutualism include, fungus-growing ants employing agricultural symbiosis, bacteria living in the guts of insects and other organisms, the fig wasp and yucca moth pollination complex, lichens with fungi and photosynthetic algae, and corals with photosynthetic algae. Nevertheless, many organisms exploit host rewards without reciprocating and thus have been branded with a myriad of not-very-flattering names such as 'cheaters', 'exploiters', 'robbers', and 'thieves'. Although cheaters impose several host cots (e.g., via damage to their reproductive organs or propagules, denying the services of a beneficial partner), their net effect on host fitness is not necessarily negative and, thus, becomes difficult to forecast.

Biogeography

The word *biogeography* is an amalgamation of *biology* and *geography*. Biogeography is the comparative study of the geographic distribution of organisms and the corresponding evolution of their traits in space and time. The Journal of Biogeography was established in 1974. Biogeography and ecology share many of their disciplinary roots. For example, the theory of island biogeography, published by the mathematician Robert MacArthur and ecologist Edward O. Wilson in 1967 is considered one of the fundamentals of ecological theory. Biogeography has a long history in the natural sciences where questions arise concerning the spatial distribution of plants and animals. Ecology and evolution provide the explanatory context for biogeographical studies. Biogeographical patterns result from ecological processes that influence range distributions, such as migration and dispersal, and from historical processes that split populations or species into different areas. The biogeographic processes that result in the natural splitting of species explains much of the modern distribution of the Earth's biota. The splitting of lineages in a species is called vicariance biogeography and it is a sub-discipline of biogeography. There are also practical applications in the field of biogeography concerning ecological systems and processes. For example, the range and distribution of biodiversity and invasive species responding to climate change is a serious concern and active area of research in context of global warming.

r/K-*Selection Theory*

A population ecology concept (introduced in MacArthur and Wilson's (1967) book, *The Theory of Island Biogeography*) is r/K selection theory, one of the first predictive models in ecology used to explain life-history evolution. The premise behind the r/K selection model is that natural selection pressures change according to population density. For example, when an island is first colonised, density of individuals is low. The initial increase in population size is *not* limited by competition, leaving an abundance of available resources for rapid population growth. These early phases of population growth experience *density-independent* forces of natural selection, which is called *r*-selection. As the population becomes more crowded, it approaches the island's carrying capacity, thus forcing individuals to compete more heavily for fewer available resources. Under crowded conditions the population experiences density-dependent forces of natural selection, called *K*-selection.

In the r/K-selection model, the first variable r is the intrinsic rate of natural increase in population size and the second variable K is the carrying capacity of a population. Different species evolve different life-history strategies spanning a continuum between these two selective forces. An r-selected species is one that has high birth rates, low levels of parental investment, and high rates of mortality before individuals reach maturity. Evolution favours high rates of fecundity in r-selected species. Many kinds of insects and invasive species exhibit r-selected characteristics. In contrast, a K-selected species has low rates of fecundity, high levels of parental investment in the young, and low rates of mortality as individuals mature. Humans and elephants are examples of species exhibiting K-selected characteristics, including longevity and efficiency in the conversion of more resources into fewer offspring.

Molecular Ecology

The important relationship between ecology and genetic inheritance predates modern techniques for molecular analysis. Molecular ecological research became more feasible with the development of rapid and accessible genetic technologies, such as the polymerase chain reaction (PCR). The rise of molecular technologies and influx of research questions into this new ecological field resulted in the publication *Molecular Ecology* in 1992. Molecular ecology uses various analytical techniques to study genes in an evolutionary and ecological context. In 1994, John Avise also played a leading role in this area of science with the publication of his book, *Molecular Markers, Natural History and Evolution*. Newer technologies opened a wave of genetic analysis into organisms once difficult to study from an ecological or evolutionary standpoint, such as bacteria, fungi and nematodes. Molecular ecology engendered a new research paradigm for investigating ecological questions considered otherwise intractable. Molecular investigations revealed previously obscured details in the tiny intricacies of nature and improved resolution into probing questions about behavioural and biogeographical ecology. For example, molecular ecology revealed promiscuous sexual behaviour and multiple male partners in tree swallows previously thought to be socially monogamous. In a biogeographical context, the marriage between genetics, ecology and evolution resulted in a new sub-discipline called phylogeography.

Human Ecology

Human ecology is the interdisciplinary investigation into the ecology of our species. "Human ecology may be defined: (1) from a bio-ecological standpoint as the study of man as the ecological dominant in plant and animal communities and systems; (2) from a bio-ecological standpoint as simply another animal affecting and being affected by his physical environment; and (3) as a human being, somehow different from animal life in general, interacting with physical and modified environments in a distinctive and creative way. A truly interdisciplinary human ecology will most likely address itself to all three." The term human ecology was formally introduced in 1921, but many sociologists, geographers, psychologists, and other disciplines were interested in human relations to natural systems centuries prior, especially in the late 19th century. Some authors have identified a new unifying science in coupled human and natural systems that builds upon, but moves beyond the field human ecology. Ecology is as much a biological science as it is a human science. "Perhaps the most important implication involves our view of human society. *Homo sapiens* is not an external disturbance, it is a keystone species within the system. In the long term, it may not be the magnitude of extracted goods and services that will determine sustainability. It may well be our disruption of ecological recovery and stability mechanisms that determines system collapse."

Relation to the Environment

The environment is dynamically interlinked, imposed upon and constrains organisms at any time throughout their life cycle. Like the term ecology, environment has different conceptual meanings and to many these terms also overlap with the concept of *nature*. Environment "...includes the physical world, the social world of human relations and the built world of human creation." The environment in ecosystems includes both physical parameters and biotic attributes. The physical environment is external to the level of biological organisation under investigation, including abiotic factors such as temperature, radiation, light, chemistry, climate and geology. The biotic environment includes genes, cells, organisms, members of the same species (conspecifics) and other species that share a habitat. The laws of thermodynamics applies to ecology by means of its physical state. Armed with an understanding of metabolic and thermodynamic principles a complete accounting of energy and material flow can be traced through an ecosystem.

Environmental and ecological relations are studied through reference to conceptually manageable and isolated parts. Once the effective environmental components are understood they conceptually link back together as a *holocoenotic* system. In other words, the organism and the environment form a dynamic whole (or umwelt). Change in one ecological or environmental factor can concurrently affect the dynamic state of an entire ecosystem.

Disturbance and Resilience

Ecosystems are regularly confronted with natural environmental variations and disturbances over time and geographic space. A disturbance is any process that removes living biomass from a community, such as a fire, flood, drought, or predation. Fluctuations causing disturbance occur over vastly different ranges in terms of magnitudes as well as distances and time periods. Disturbances, such as fire, are both cause and product of natural fluctuations in death rates, species assemblages, and biomass densities within an ecological community. These disturbances create places of renewal where new directions emerge out of the patchwork of natural experimentation and opportunity. Ecological resilience is a cornerstone theory in ecosystem management. Biodiversity fuels the resilience of ecosystems acting as a kind of regenerative insurance.

Metabolism and the Early Atmosphere

The Earth formed approximately 4.5 billion years ago and environmental conditions were too extreme for life to form for the first 500 million years. During this early Hadeanperiod, the Earth started to cool, allowing a crust and oceans to form. Environmental conditions were unsuitable for the origins of life for the first billion years after the Earth formed. The Earth's atmosphere transformed from being dominated by hydrogen, to one composed mostly of methane and ammonia. Over the next billion years the metabolic activity of life transformed the atmosphere to higher concentrations of carbon dioxide, nitrogen, and water vapour. These gases changed the way that light from the sun hit the Earth's surface and greenhouse effects trapped heat. There were untapped sources of free energy within the mixture of reducing and oxidising gasses that set the stage for primitive ecosystems to evolve and, in turn, the atmosphere also evolved.

The leaf is the primary site ofphotosynthesis in most plants.

Throughout history, the Earth's atmosphere and biogeochemical cycles have been in a dynamic equilibrium with planetary ecosystems. The history is characterised by periods of significant transformation followed by millions of years of stability. The evolution of the earliest organisms, likely anaerobic methanogen microbes, started the process by converting atmospheric hydrogen into methane ($4H_2 + CO_2$ '! $CH_4 + 2H_2O$). Anoxygenic photosynthesis converting hydrogen sulfide into other sulfur compounds or water (for example $2H_2S + CO_2 + h\upsilon$ '! $CH_2O + H_2O + 2S$), as occurs in deep sea hydrothermal vents today, reduced hydrogen concentrations and increased atmospheric methane. Early forms of fermentation also increased levels of atmospheric methane. The transition to an oxygen dominant atmosphere (the *Great Oxidation*) did not begin until approximately 2.4-2.3 billion years ago, but photosynthetic processes started 0.3 to 1 billion years prior.

Radiation: Heat, Temperature and Light

The biology of life operates within a certain range of temperatures. Heat is a form of energy that regulates temperature. Heat affects growth rates, activity, behaviour and primary production. Temperature is largely dependent on the incidence of solar radiation. The latitudinal and longitudinal spatial variation of temperature greatly affects climates and consequently the distribution of biodiversity and levels of primary production in different ecosystems or biomes across the planet. Heat and temperature relate importantly to metabolic activity. Poikilotherms, for example, have a body temperature that is largely regulated and dependent on the temperature of the external environment. In contrast, homeotherms regulate their internal body temperature by expending metabolic energy.

There is a relationship between light, primary production, and ecological energy budgets. Sunlight is the primary input of energy into the planet's ecosystems. Light is composed of electromagnetic energy of different wavelengths. Radiant energy from the sun generates heat, provides photons of light measured as active energy in the chemical reactions of life, and also acts as a catalyst for genetic mutation. Plants, algae, and some bacteria absorb light and assimilate the energy through photosynthesis. Organisms capable of assimilating energy by photosynthesis or through inorganic fixation of H_2S are autotrophs. Autotrophs—responsible for primary production— assimilate light energy that becomes metabolically stored as potential energy in the form of biochemical enthalpic bonds.

Physical Environments

Water

The rate of diffusion of carbon dioxide and oxygen is approximately 10,000 times slower in water than it is in air. When soils become flooded, they quickly lose oxygen and transform into a low-concentration (hypoxic- O_2 concentration lower than 2 mg/liter) environment and eventually become completely (anoxic) environment where anaerobic bacteria thrive among the roots. Water also influences the spectral composition and amount of light as it reflects off the water surface and submerged particles. Aquatic plants exhibit a wide variety of morphological and physiological adaptations that allow them to survive, compete and diversify these environments. For example, the roots and stems develop large air spaces (Aerenchyma) that regulate the efficient transportation gases (for example, CO_2 and O_2) used in respiration and photosynthesis. In drained soil, microorganisms use oxygen during respiration. In aquatic environments, anaerobic soil microorganisms use nitrate, manganese ions, ferric ions, sulfate, carbon dioxide and some organic compounds. The activity of soil microorganisms and the chemistry of the water reduces the oxidation-reduction potentials of the water. Carbon dioxide, for example, is reduced to methane (CH_4) by methanogenic bacteria. Salt water plants (or halophytes) have specialised physiological adaptations, such as the development of special organs for shedding salt and osmo-regulate their internal salt (NaCl) concentrations, to live in estuarine, brackish, or oceanic environments. The physiology of fish is also specially adapted to deal with high levels of salt through osmoregulation. Their gills form electrochemical gradients that mediate salt excresion in saline environments and uptake in fresh water.

Gravity

The shape and energy of the land is affected to a large degree by gravitational forces. On a larger scale, the distribution of gravitational forces on the earth are uneven and influence the shape and movement of tectonic plates as well as having an influence on geomorphic processes such as orogeny and erosion. These forces govern many of the geophysical properties and distributions of ecological biomes across the Earth. On a organism scale, gravitational forces provide directional cues for plant and fungal growth (gravitropism), orientation cues for animal migrations, and influence the biomechanics and size of animals. Ecological traits, such as allocation of biomass in trees during growth are subject to mechanical failure as gravitational forces influence the position and structure of branches and leaves. The cardiovascular systems of all animals are functionally adapted to overcome pressure and gravitational forces that change according to the features of organisms (e.g., height, size, shape), their behaviour (e.g., diving, running, flying), and the habitat occupied (e.g., water, hot deserts, cold tundra).

Pressure

Climatic and osmotic pressure places physiological constraints on organisms, such as flight and respiration at high altitudes, or diving to deep ocean depths. These constraints influence vertical limits of ecosystems in the biosphere as organisms are physiologically sensitive and adapted to atmospheric and osmotic water pressure differences. Oxygen levels, for example, decrease with increasing pressure and are a limiting factor for life at higher altitudes. Watertransportation through trees is another important ecophysiological parameter where osmotic pressure gradients factor in. Water pressure in the depths of oceans requires that organisms adapt to these conditions. For example, mammals, such as whales, dolphins and seals are specially adapted to deal with changes in sound due to water pressure differences. Different species of hagfish provide another example of adaptation to deep-sea pressure through specialised protein adaptations.

Wind and Turbulence

Turbulent forces in air and water have significant effects on the environment and ecosystem distribution, form and dynamics. On a planetary scale, ecosystems are affected by circulation patterns in the global trade winds. Wind power and the turbulent forces it creates can influence heat, nutrient, and biochemical profiles of ecosystems. For example, wind running over the surface of a lake creates turbulence, mixing the water column and influencing the environmental profile to create thermally layered zones, partially governing how fish, algae, and other parts of the aquatic ecology are structured. Wind speed and turbulence also exert influence on rates of evapotranspiration rates and energy budgets in plants and animals. Wind speed, temperature and moisture content can vary as winds travel across different land features and elevations. The westerlies, for example, come into contact with the coastal and interior mountains of western North America to produce a rain shadow on the leeward side of the mountain. The air expands and moisture condenses as the winds move up in elevation which can cause precipitation; this is called orographic lift. This environmental process produces spatial divisions in biodiversity, as species adapted to wetter conditions are range-restricted to the coastal mountain valleys and unable to migrate across the xeric ecosystems of the Columbia Basin to intermix with sister lineages that are segregated to the interior mountain systems.

Fire

Plants convert carbon dioxide into biomass and emit oxygen into the atmosphere. Approximately 350 million years ago (near the

Devonian period) the photosynthetic process brought the concentration of atmospheric oxygen above 17%, which allowed combustion to occur. Fire releases CO_2 and converts fuel into ash and tar. Fire is a significant ecological parameter that raises many issues pertaining to its control and suppression in management. While the issue of fire in relation to ecology and plants has been recognised for a long time, Charles Cooper brought attention to the issue of forest fires in relation to the ecology of forest fire suppression and management in the 1960s.

Fire creates environmental mosaics and a patchiness to ecosystem age and canopy structure. Native North Americans were among the first to influence fire regimes by controlling their spread near their homes or by lighting fires to stimulate the production of herbaceous foods and basketry materials. The altered state of soil nutrient supply and cleared canopy structure also opens new ecological niches for seedling establishment. Most ecosystem are adapted to natural fire cycles. Plants, for example, are equipped with a variety of adaptations to deal with forest fires. Some species (e.g., *Pinus halepensis*) cannot germinate until after their seeds have lived through a fire. This environmental trigger for seedlings is called serotiny. Some compounds from smoke also promote seed germination. Fire plays a major role in the persistence and resilience of ecosystems.

Biogeochemistry

Ecologists study and measure nutrient budgets to understand how these materials are regulated, flow, and recycled through the environment. This research has led to an understanding that there is a global feedback between ecosystems and the physical parameters of this planet including minerals, soil, pH, ions, water and atmospheric gases. There are six major elements, including H (hydrogen), C (carbon), N (nitrogen), O (oxygen), S (sulfur), and P (phosphorus) that form the constitution of all biological macromolecules and feed into the Earth's geochemical processes. From the smallest scale of biology the combined effect of billions upon billions of ecological processes amplify and ultimately regulate the biogeochemical cycles of the Earth. Understanding the relations and cycles mediated between these elements and their ecological pathways has significant bearing toward understanding global biogeochemistry.

The ecology of global carbon budgets gives one example of the linkage between biodiversity and biogeochemistry. For starters, the Earth's oceans are estimated to hold 40,000 gigatonnes (Gt) carbon, vegetation and soil is estimated to hold 2070 Gt carbon, and fossil fuel emissions are estimated to emit an annual flux of 6.3 Gt carbon. At different times in the Earth's history there has been major restructuring in these global carbon budgets that was regulated to a large extent by the ecology of the land. For example, through the early-mid Eocene volcanic outgassing, the oxidation of methane stored in wetlands, and seafloor gases increased atmospheric CO_2 (carbon dioxide) concentrations to levels as high as 3500 ppm.

In the Oligocene, from 25 to 32 million years ago, there was another significant restructuring in the global carbon cycle as grasses evolved a special type of C4 photosynthesis and expanded their ranges. This new photosynthetic pathway evolved in response to the drop in atmospheric CO_2 concentrations below 550 ppm. These kinds of ecosystem functions feed back significantly into global atmospheric models for carbon cycling. Loss in the abundance and distribution of biodiversity causes global carbon cycle feedbacks that are expected to increase rates of global warming in the next century. The effect of global warming melting large sections of permafrost creates a new mosaic of flooded areas where decomposition results in the emission of methane (CH_4). Hence, there is a relationship between global warming, decomposition and respiration in soils and wetlands producing significant climate feedbacks and altered global biogeochemical cycles. There is concern over increases in atmospheric methane in the context of the global carbon cycle, because methane is also a greenhouse gas that is 23 times more effective at absorbing long-wave radiation than CO_2 on a 100 year time scale.

History

Early Beginnings

Ecology has a complex origin due in large part to its interdisciplinary nature. Ancient philosophers of Greece, including Hippocrates and Aristotle were among the first to record their observations on natural history. However, philosophers in ancient Greece viewed life as a static element that did not require an understanding of adaptation, a modern cornerstone of ecological theory. Topics more familiar in the modern context, including food chains, population regulation, and productivity, did not develop until the 1700s through the published works of microscopist Antoni van Leeuwenhoek (1632–1723) and botanist Richard Bradley(1688?-1732). Biogeographer Alexander von Humbolt (1769–1859) was another early pioneer in ecological thinking and was among the first to recognise ecological gradients. Humbolt alluded to the modern ecological law of species to area relationships.

In the early 20th century, ecology was an analytical form of natural history. Following in the traditions of Aristotle, the descriptive nature of natural history examined the interaction of organisms with both their environment and their community. Natural historians, including James Hutton and Jean-Baptiste Lamarck, contributed significant works that laid the foundations of the modern ecological sciences. The term "ecology" (German: *Oekologie*) is of a more recent origin and was first coined by the German biologist Ernst Haeckel in his book *Generelle Morphologie der Organismen* (1866). Haeckel was a zoologist, artist, writer, and later in life a professor of comparative anatomy.

Opinions differ on who was the founder of modern ecological theory. Some mark Haeckel's definition as the beginning, others say it was Eugenius Warming with the writing of Oecology of Plants: An Introduction to the Study of Plant Communities(1895). Ecology may also be thought to have begun with Carl Linnaeus' research principals on the economy of nature that matured in the early 18th century. He founded an early branch of ecological study he called the economy of nature. The works of Linnaeus influenced Darwin in *The Origin of Species* where he adopted the usage of Linnaeus' phrase on the *economy or polity of nature*. Linnaeus was the first to frame the balance of nature as a testable hypothesis. Haeckel, who admired Darwin's work, defined ecology in reference to the economy of nature which has led some to question if ecology is synonymous with Linnaeus' concepts for the economy of nature.

The modern synthesis of ecology is a young science, which first attracted substantial formal attention at the end of the 19th century (around the same time as evolutionary studies) and become even more popular during the 1960s environmental movement, though many observations, interpretations and discoveries relating to ecology extend back to much earlier studies in natural history. For example, the concept on the balance or regulation of nature can be traced back to Herodotos (died *c.* 425 BC) who described an early account of mutualism along the Nile river where crocodiles open their mouths to beneficially allow sandpipers safe access to remove leeches. In the broader contributions to the historical development of the ecological sciences, Aristotle is considered one of the earliest naturalists who had an influential role in the philosophical development of ecological sciences. One of Aristotle's students, Theophrastus, made astute ecological observations about plants and posited a philosophical stance about the autonomous relations between plants and their environment that is more in line with modern ecological thought. Both Aristotle and

Theophrastus made extensive observations on plant and animal migrations, biogeography, physiology, and their habits in what might be considered an analogue of the modern ecological niche. Hippocrates, another Greek philosopher, is also credited with reference to ecological topics in its earliest developments.

From Aristotle to Darwin the natural world was predominantly considered static and unchanged since its original creation. Prior to *The Origin of Species* there was little appreciation or understanding of the dynamic and reciprocal relations between organisms, their adaptations and their modifications to the environment. While Charles Darwin is most notable for his treatise on evolution, he is also one of the founders of soil ecology. In *The Origin of Species* Darwin also made note of the first ecological experiment that was published in 1816. In the science leading up to Darwin the notion of evolving species was gaining popular support. This scientific paradigm changed the way that researchers approached the ecological sciences.

After the turn of 20th Century

Some suggest that the first ecological text (*Natural History of Selborne*) was published in 1789, by Gilbert White (1720–1793). The first American ecology book was published in 1905 by Frederic Clements. In his book, Clements forwarded the idea of plant communities as a superorganism. This publication launched a debate between ecological holism and individualism that lasted until the 1970s. The Clements superorganism concept proposed that ecosystems progress through regular and determined stages of seral development that are analogous to developmental stages of an organism whose parts function to maintain the integrity of the whole. The Clementsian paradigm was challenged by Henry Gleason. According to Gleason, ecological communities develop from the unique and coincidental association of individual organisms. This perceptual shift placed the focus back onto the life histories of individual organisms and how this relates to the development of community associations.

The Clementsian superorganism theory has not been completely rejected, but some suggest it was an overextended application of holism. Holism remains a critical part of the theoretical foundation in contemporary ecological studies. Holism was first introduced in 1926 by a polarizing historical figure, a South African General named Jan Christian Smuts. Smuts was inspired by Clement's superorganism theory as he developed and published on the concept of holism, which contrasts starkly against his racial political views as the father of apartheid. Around the same time, Charles Elton pioneered the concept of food chains in his classical book "Animal Ecology". Elton defined ecological relations using concepts of food chains, food cycles, food size, and described numerical relations among different functional groups and their relative abundance. Elton's 'food cycle' was replaced by 'food web' in a subsequent ecological text.

Ecology has developers in many nations, including Russia's Vladimir Vernadsky and his founding of the biosphere concept in the 1920s or Japan's Kinji Imanishi and his concepts of harmony in nature and habitat segregation in the 1950s. The scientific recognition or importance of contributions to ecology from other cultures is hampered by language and translation barriers.

Chapter 5

Habitat Conservation

Habitat conservation is a land management practice that seeks to conserve, protect and restore, habitat areas for wild plants and animals, especially conservation reliant species, and prevent their extinction, fragmentation or reduction in range. It is a priority of many groups that cannot be easily characterised in terms of any one ideology.

Values of Natural Habitat

The natural environment is a source for a wide range of resources that can be exploited for economic profit, for example timber is harvested from forests and clean water is obtained from natural streams. However, land development from anthropogenic economic growth often causes a decline in the ecological integrity of nearby natural habitat. For instance, this was an issue in the northern rocky mountains of the USA.

However, there is also economic value in conserving natural habitat. Financial profit can be made from tourist revenue, particularly in the tropics where species diversity is high. The cost of repairing damaged ecosystems is considered to be much higher than the cost of conserving natural ecosystems.

Measuring the worth of conserving different habitat areas is often criticized as being too utilitarian from a philosophical point of view.

Food Security

Habitat conservation is important in maintaining biodiversity, an essential part of global food security. There is evidence to support a trend of accelerating erosion of the genetic resources of agricultural plants and animals. An increase in genetic similarity of agricultural plants and animals means an increased risk of food loss from major epidemics. Wild species of agricultural plants have been found to be more resistant to disease, for example the wild corn species Teosinte is resistant to 4 corn diseases that affect human grown crops. A combination of seed banking and habitat conservation has been proposed to maintain plant diversity for food security purposes.

Classifying Environmental Values

Pearce and Moran outlined the following method for classifying environmental uses:

- Direct extractive uses: e.g. timber from forests, food from plants and animals
- Indirect uses: e.g. ecosystem services like flood control, pest control, erosion protection
- Optional uses: future possibilities e.g. unknown but potential use of plants in chemistry/medicine
- Non-use values:

Bequest value (benefit of an individual who knows that others may benefit from it in future)

Passive use value (sympathy for natural environment, enjoyment of the mere existence of a particular species)

Impacts

Natural Causes

Habitat loss and destruction can occur both naturally and through anthropogenic causes. Events leading to natural habitat loss include climate change, catastrophic events such as volcanic explosions and through the interactions of invasive and non-invasive species. Natural climate change, events have previously been the cause of many widespread and large scale losses in habitat. For example, some of the mass extinction events generally referred to as the "Big Five" have coincided with large scale such as the Earth entering an ice age, or alternate warming events. Other events in the big five also have their roots in natural causes, such as volcanic explosions and meteor collisions.

The Chicxulub impact is one such example, which has previously caused widespread losses in habitat as the Earth either received less sunlight or grew colder, causing certain fauna and flora to flourish whilst others perished. Previously known warm areas in the tropics, the most sensitive habitats on Earth, grew colder, and areas such as Australia developed radically different flora and fauna to those seen today. The big five mass extinction events have also been linked to sea level changes, indicating that large scale marine species loss was strongly influenced by loss in marine habitats, particularly shelf habitats. Methane-driven oceanic eruptions have also been shown to have caused smaller mass extinction events.

Human Impacts

Since radiating out from its birthplace in Africa, Homo sapiens has been the cause of many species' extinction. Due to humans' propensity to shape and modify their environment, the habitat of other species often become altered or destroyed as a result of human actions. Even before the modern industrial era, humans were having widespread, catastrophic effects on the environment. A good example of this is found in Aboriginal Australians and Australian megafauna. Aboriginal hunting practices, which included burning large sections of forest at a time, eventually altered and changed Australia's vegetation so much that many herbivorous megafauna species were left with no habitat and were driven into extinction. Once herbivorous megafauna species became extinct, carnivorous megafauna species soon followed. In the recent past, humans have been responsible for causing more extinctions within a given period of time than ever before deforestation, pollution, anthropogenic climate change and human settlements have all been driving forces in altering or destroying habitats. The destruction of ecosystems such as rainforests has resulted in countless habitats being destroyed. These biodiversity hotspots are home to millions of habitat specialists, which do not exist beyond a tiny area. Once their habitat is destroyed, they cease to exist. This destruction has a follow-on effect, as species which coexist or depend upon the existence of other species also become extinct, eventually resulting in the collapse of an entire ecosystem. These time-delayed extinctions are referred to as the extinction debt, which is the result of destroying and fragmenting habitats. As a result of anthropogenic modification of the environment, the extinction rate has climbed to the point where the Earth is now within a sixth mass extinction event, as commonly agreed by biologists. This has been particularly evident, for example, in the rapid decline in the number of amphibian species worldwide.

Approaches and Methods of Habitat Conservation

Determining the size, type and location of habitat to conserve is a complex area of conservation biology. Although difficult to measure and predict, the conservation value of a habitat is often a reflection of the quality (e.g. species abundance and diversity), endangerment of encompassing ecosystems, and spatial distribution of that habitat.

Identifying Priority Habitats for Conservation

Habitat conservation is vital for protecting species and ecological processes. It is important to conserve and protect the space/area in which that species occupies. Therefore, areas classified as 'biodiversity hotspots', or those in which a flagship, umbrella, or endangered species inhabits are often the habitats that are given precedence over others. Species that possess an elevated risk of extinction are given the highest priority and as a result of conserving their habitat, other species in that community are protected thus serving as an element of gap analysis. In the United States of America, a Habitat Conservation Plan (HCP) is often developed to conserve the environment in which a specific species inhabits. Under the U.S. Endangered Species Act (ESA) the habitat that requires protection in an HCP is referred to as the 'critical habitat'. Multiple-species HCPs are becoming more favourable then single-species HCPs as they can potentially protect an array of species before they warrant listing under the ESA, as well as being able to conserve broad ecosystem components and processes. As of January 2007, 484 HCPs were permitted across the United States, 40 of which covered 10 or more species. The San Diego Multiple Species Conservation Plan (MSCP) encompasses 85 species in a total area of 26,000-km^2. Its aim is to protect the habitats of multiple species and overall biodiversity by minimising development in sensitive areas.

HCPs require clearly defined goals and objectives, efficient monitoring programs, as well as successful communication and collaboration with stakeholders and land owners in the area. Reserve design is also important and requires a high level of planning and management in order to achieve the goals of the HCP. Successful reserve design often takes the form of a hierarchical system with the most valued habitats requiring high protection being surrounded by buffer habitats that have a lower protection status. Like HCPs, hierarchical reserve design is a method most often used to protect a single species, and as a result habitat corridors are maintained, edge effects are reduced and a broader suite of species are protected.

How Much Habitat is Enough

A range of methods and models currently exist that can be used to determine how much habitat is to be conserved in order to sustain a viable population. Modelling tools often rely on the spatial scale of the area as an indicator of conservation value. There has been an increase in emphasis on conserving few large areas of habitat as opposed to many small areas. This idea is often referred to as the "single large or several small", SLOSS debate, and is a highly controversial area among conservation biologists and ecologists. The reasons behind the argument that "larger is better" include the reduction in the negative impacts of patch edge effects, the general idea that species richness increases with habitat area and the ability of larger habitats to support greater populations with lower extinction probabilities. Noss & Cooperrider support the "larger is better" claim and developed a model that implies areas of habitat less than 1000ha are "tiny" and of low conservation value. However, Shwartz suggests that although "larger is better", this does not imply that "small is bad". Shwartz argues that human induced habitat loss leaves no alternative to conserving small areas. Furthermore, he suggests many endangered species which are of high conservation value, may only be restricted to small isolated patches of habitat, and thus would be overlooked if larger areas were given a higher priority. The shift to conserving larger areas is somewhat justified in society by placing more value on larger vertebrate species, which naturally have larger habitat requirements.

The Conservation Movement

For much of human history, nature had been seen as a resource, one that could be controlled and used for personal and economic gain. The idea was that plants only existed to feed animals and animals only existed to feed man. The land itself had limited value only extending to the resources it could provide such as minerals and oil. Throughout the 18th and 19th centuries social views started to change and when in 1872, the world's first national park, Yellowstone National Park in the United States of America, was declared, did the official conservation movement begin.

Rather than focusing on the economic or material benefits associated with nature, humans began to appreciate the value in the nature itself and the need to protect pristine wilderness. By the middle of the 20th century countries such as the United States, Canada, and Britain understood this appreciation and instigated laws and legislation in order to ensure that the most fragile and beautiful environments would be protected for generations to come. Today with the help of NGO's, not-for profit organisations and governments worldwide there is a stronger movement taking place, with a deeper understanding of habitat conservation with the aim of protecting delicate habitats and preserving biodiversity on a global scale. The commitment and actions of small volunteering association in villages and towns, that endeavour to emulate the work done by well known

Conservation Organisations, is paramount in ensuring generations that follow understand the importance of conserving natural resources. A village conservation group with the mission statement "We are committed to protecting and enhancing the natural environment in and around the adjoining villages of Ouston and Urpeth." may one day inspire a child who becomes the employee of a worldwide conservation organisation.

Examples of Current Conservation Organisations

The Nature Conservancy

Since its formation in 1951 The Nature Conservancy has slowly developed into one of the world's largest conservation organisations. Currently operating in over 30 countries, across 5 continents worldwide, The Nature Conservancy aims to protect nature and its assets for future generations. The organisation purchases land or accepts land donations with the intension of conserving its natural resources. In 1955 The Nature Conservancy purchased its first 60-acre plot near the New York/Connecticut border in the United States of America. Today the Conservancy has expanded to protect over 119 million acres of land, 5,000 river miles as well as participating in over 1000 marine protection programs across the globe. Since its beginnings The Nature Conservancy has understood the benefit in taking a scientific approach towards habitat conservation. For the last decade the organisation has been using a collaborative, scientific method known as 'Conservation by Design'. By collecting and analysing scientific data The Conservancy is able to holistically approach the protection of various ecosystems. This process determines the habitats that need protection, specific elements that should be conserved as well as monitoring progress so more efficient practices can be developed for the future.

The Nature Conservancy currently has a large number of diverse projects in operation. They work with countries around the world to protect forests, river systems, oceans, deserts and grasslands. In all cases the aim is to provide a sustainable environment for both the plant and animal life forms that depend on them as well as all future generations to come.

World Wildlife Fund (WWF)

The WWF was first formed in 1961 after a group of passionate conservationists signed what is now referred to as the Morges Manifesto. Fifty years on, WWF is currently operating in over 100 countries across 5 continents with a current listing of over 5 million supporters. One of the first projects of WWF was assisting in the creation of the Charles Darwin Research Foundation which aided in the protection of diverse range of unique species existing on the Galápagos' Islands, Ecuador.

It was also a WWF grant that helped with the formation of the College of African Wildlife Management in Tanzania which today focuses on teaching a wide range of protected area management skills in areas such as ecology, range management and law enforcement. The WWF has since gone on to aid in the protection of land in Spain, creating the Coto Doñana National Park in order to conserve migratory birds and The Democratic Republic of Congo, home to the world's largest protected wetlands.

The WWF also initiated a debt-for-nature concept which allows the country to put funds normally allocated to paying off national debt, into conservation programs that protect its natural landscapes. Countries currently participating include Madagascar, the first country to participate which since 1989 has generated over $US50 million towards preservation, Bolivia, Costa Rica, Ecuador, Gabon, the Philippines and Zambia.

Rare Conservation

Rare has been in operation since 1973 with current global partners in over 50 countries and offices in the United States of America, Mexico, the Philippines, China and Indonesia. Rare focuses on the human activity that threatens biodiversity and habitats such as overfishing and unsustainable agriculture. By engaging local communities and changing behaviour Rare has been able to launch campaigns to protect areas in most need of conservation. The key aspect of Rare's methodology is their "Pride Campaign's". For example, in the Andes in South America, Rare has partnered with 11 different sites with the intention of creating incentives to develop watershed protection practices. In the Southeast Asia's "coral triangle" Rare is training fishers in local communities to better manage the areas around the coral reefs in order to lessen human impact. Such programs last for three years with the aim of changing community attitudes so as to conserve fragile habitats and provide ecological protection for years to come.

WWF Netherlands

WWF Netherlands, along with ARK Nature, Wild Wonders of Europe and Conservation Capital have started the Rewilding Europe project. This project intents to rewild several areas in Europe.

List of Environmental Organisations

Intergovernmental Organisations

Worldwide:

- Intergovernmental Panel on Climate Change (IPCC)
- United Nations Environment Programme (UNEP)
- Earth System Governance Project

Regional:

- European Environment Agency (EEA)
- Partnerships in Environmental Management for the Seas of East Asia (PEMSEA)

Local Governments:

- ICLEI.

Government Organisations

The governments of many countries have ministries or agencies devoted to monitoring and protecting the environment:

Brazil

- IBAMA.

Canada

- Environment Canada.

China

- Ministry of Environmental Protection.

Denmark

- Danish Ministry of Climate and Energy.

Germany

- Federal Ministry for Environment, Nature Conservation and Nuclear Safety.

Hong Kong

- Environmental Protection Department.

India

- Gujarat Pollution Control Board
- Ministry of Environment and Forests
- The Earth Organisation.

Indonesia
- Directorate General of Forest Protection and Nature Conservation.

Ireland
- Environmental Protection Agency.

Isle of Man
- Manx National Trust.

Israel
- Ministry of the Environment.

Mexico
- Secretariat of the Environment and Natural Resources.

Netherlands
- Ministry of Housing, Spatial Planning and the Environment.

New Zealand
- Department of Conservation.
- Ministry for the Environment.
- Parliamentary Commissioner for the Environment.

Nigeria
- Kano State Environmental Planning and Protection Agency.

Norway
- Norwegian Ministry of the Environment
- Norwegian Directorate for Nature Management
- Norwegian Pollution Control Authority.

Philippines
- Department of Environment and Natural Resources.

Portugal
- Ministry for Environment, Spatial Planning and Regional Development.

Republic of China (Taiwan)
- Environmental Protection Administration.

Saudi Arabia
- Saudi Environmental Society.

United Kingdom
- Department for Environment, Food and Rural Affairs.

England
- English Heritage.
- Environment Agency.
- Natural England.

Scotland
- Historic Scotland
- Scottish Natural Heritage
- Scottish Environment Protection Agency.

Wales
- Cadw
- Countryside Council for Wales
- Environment Agency Wales.

Northern Ireland
- Northern Ireland Environment Agency.

United States
- United States Environmental Protection Agency
- United States Fish and Wildlife Service
- United States National Park Service.

Native American Nations
- Inter-Tribal Environmental Council.

Private Organisations (Environmental NGOs)
These organisations are involved in environmental management, lobbying, advocacy, or conservation efforts:

International
- 350.org
- Anti-nuclear movement
- Antinea Foundation
- A Rocha
- Biofuelwatch
- Biosphere Expeditions
- Bioversity International

- BirdLife International
- Confederation of European Environmental Engineering Societies
- Conservation International
- Earth Charter Initiative
- Forests and the European Union Resource Network (FERN)
- Fauna and Flora International
- Forest Stewardship Council
- Friends of Nature
- Friends of the Earth
- Gaia Mater (the mother Earth)
- Global Footprint Network
- Global Witness
- Great Transition Initiative
- Green Actors of West Africa (GAWA)
- Green Cross International
- Greenpeace
- Interamerican Association for Environmental Defence
- International Union for Conservation of Nature (IUCN)
- International Analogue Forestry Network
- International Network for Sustainable Energy (INFORSE)
- The Mountain Institute
- Mountain Wilderness
- NatureServe
- Plant A Tree Today Foundation (PATT)
- Project AWARE
- Rainforest Alliance
- Sandwatch
- Seeds of Survival of USC Canada
- Society for the Environment (SocEnv)
- Taiga Rescue Network (TRN)
- The Climate Project
- The Resource Foundation
- Wetlands International

- Wildlife Conservation Society
- Wolf Preservation Foundation (WPF)
- World Business Council for Sustainable Development
- Worldchanging
- World Conservation Union (WCN)
- World Land Trust(WLT)
- World Resources Institute (WRI)
- World Union for Protection of Life (WUPL)
- Worldwatch Institute
- World Wide Fund for Nature (WWF)
- Xerces Society
- Yellowstone to Yukon Conservation Initiative.

Regional

Africa

- Environmental Foundation for Africa.

Europe

- Climate Action Network- Europe (CAN-Europe)
- European Association of Environmental and Resource Economists (EAERE)
- European Biomass Association
- European Environmental Bureau (EEB)
- INFORSE-Europe
- Coastwatch Europe.

North America

- Fund for Wild Nature
- Green Zionist Alliance
- International Joint Commission
- North American Native Fishes Association
- Rivers Without Borders
- Wild Farm Alliance.

South America

- Amazon Watch
- Rainforest Foundation Fund.

National

Australia

- Australian Conservation Foundation
- Australian Koala Foundation
- Australian Network of Environmental Defenders Offices
- Australian Student Environment Network
- Australian Wildlife Conservancy
- Banksia Environmental Foundation
- Birds Australia
- Blue Wedges
- Clean Ocean Foundation
- Environment Victoria
- Greening Australia
- Landcare Australia
- Public Transport Users Association
- The Wilderness Society (Australia)
- Wildlife Watch Australia.

Category:Environmental organisations based in Australia.

Austria

- Transitforum Austria Tirol.

Bahamas

- Friends of the Environment.

Bolivia

- Comunidad Inti Wara Yassi (CIWY).

Canada

- Ancient Forest Alliance
- Bird Protection Quebec
- Canadian Association of Physicians for the Environment
- Canadian Environmental Law Association
- Canadian Environmental Network
- Canadian Youth Climate Coalition
- Canadian Parks and Wilderness Society
- David Suzuki Foundation

- Ecojustice Canada
- Earth Liberation Army (ELA)
- Earth Rangers
- Energy Probe
- Green Action Centre
- Manitoba Eco-Network
- Nature Canada
- Pembina Institute
- Regenesis (non-profit organisation)
- Sierra Youth Coalition
- The Society for the Preservation of Wild Culture
- Toronto Environmental Alliance (TEA)
- Western Canada Wilderness Committee.

Category:Environmental organisations based in Canada.

China

- Friends of Nature (China)
- Green Camel Bell.

Croatia

- Ekoloko društvo Zeleni Osijek.

Czech Republic

- Environmental Law Service (ELS)
- Hnutí DUHA- Friends of the Earth Czech Republic.

Denmark

- Danish Organisation for Renewable Energy (OVE).

Germany

- Bund für Umwelt und Naturschutz Deutschland (BUND) = Friends of the Earth, Germany
- BUNDjugend (BUND's Youth organisation)
- EarthLink e.V.
- Fuck for Forest
- Naturschutzbund Deutschland (NABU)- German Nature Conservation Society
- Robin Wood
- Zeitz Foundation.

Greece
- Environmental Centre ARCTUROS.

Hong Kong
- Clear the Air (Hong Kong)
- Friends of the Earth (HK)
- Green Power
- Lights Out Hong Kong
- Society for Protection of the Harbour
- The Conservancy Association.

India
- Awaaz Foundation
- CERE India
- Conserve
- Exnora International
- Foundation for Ecological Security
- Goa Foundation
- Centre for Science and Environment.

Indonesia
- Borneo Orangutan Survival Foundation.

Ireland
- Gluaiseacht
- Irish Peatland Conservation Council (IPCC).

Israel
- Israel Union for Environmental Defence (IUED)
- Green Movement
- Palestinian Environmental NGOs Network
- Society for the Protection of Nature in Israel (SPNI)
- Zalul Environmental Association.

Kenya
- Green Belt Movement.

Macedonia
- Macedonian Ecological Society.

Madagascar
- L'Homme et L'Environnement.

Malta
- BirdLife Malta.

Mexico

Nepal
- International Centre for Integrated Mountain Development
- National Trust for Nature Conservation.

Netherlands
- Milieudefensie.

New Zealand
- Buller Conservation Group
- Conservation Volunteers New Zealand
- Environment and Conservation Organisations of Aotearoa New Zealand (ECO)
- Native Forest Restoration Trust
- New Zealand Ecological Restoration Network
- New Zealand Institute of Environmental Health (NZIEH)
- Royal Forest and Bird Protection Society of New Zealand
- Save Happy Valley Campaign
- TerraNature
- Waipoua Forest Trust.

Norway
- Bellona Foundation
- Eco-Agents
- Norwegian Society for the Conservation of Nature
- Green Warriors of Norway (*Norges Miljøvernforbund*)
- Nature and Youth
- Zero Emission Resource Organisation.

Philippines
- Greenpeace Southeast Asia- Philippines
- Haribon Foundation
- Sibuyanons Against Mining.

Portugal
- Quercus.

Sierra Leone
- Enforac (Environmental Forum for Action).

South Africa
- Cape Town Ecology Group
- Dolphin Action & Protection Group
- Earthlife Africa
- Endangered Wildlife Trust
- EThekwini ECOPEACE
- Groundwork
- Koeberg Alert
- The Earth Organisation
- Wildlife & Environment Society.

Spain
- Asociación pola defensa da ría.

Sweden
- Fältbiologerna.

Ukraine
- Ukraine Nature Conservation Society (UkrTOP).

United Kingdom
- Association for Environment Conscious Building
- Bicycology
- Campaign for Better Transport
- Campaign for National Parks (CNP)
- Campaign to Protect Rural England
- Centre for Alternative Technology (CAT)
- Chartered Institution of Water and Environmental Management (CIWEM)
- The Corner House
- Creative Environmental Networks (CEN)
- Earth Liberation Front (ELF)
- Earth Liberation Prisoner Support Network (ELPSN)

- Environmental Investigation Agency
- Environmental Justice Foundation
- Environmental Law Foundation (ELF)
- Environmental Protection UK
- Forest Peoples Programme
- Green Alliance
- Groundwork UK
- The Institution of Environmental Sciences
- Marine Conservation Society
- John Muir Trust
- People & Planet
- Plane Stupid
- RSPB (Royal Society for the Protection of Birds)
- Scottish Wildlife Trust
- Stop Climate Chaos
- The Civic Trust
- World Land Trust
- The Wildlife Trusts
- Town and Country Planning Association
- UK Environmental Law Association (UKELA)
- Whale and Dolphin Conservation Society
- Woodland Trust.

United States

- 41pounds.org
- Abalone Alliance (historic)
- Adirondack Mountain Club
- African American Environmentalist Association
- African Wild Dog Conservancy
- Albatross Foundation USA
- Allegheny Land Trust
- Alliance for Climate Protection
- Alliance to Save Energy
- American Bird Conservancy
- American Farmland Trust

- American Wind and Wildlife Institute
- Animal Protection and Rescue League (APRL)
- Appalachian Voices
- Arlington Coalition on Transportation (ACT)
- Association of Environmental Professionals
- Audubon movement
- Bonneville Environmental Foundation (BEF)
- Builders for the Bay
- Centre for a New American Dream
- Centre for International Environmental Law
- Centre for Biological Diversity
- Chesapeake Bay Foundation
- Citizens Campaign for the Environment
- Committee for a Constructive Tomorrow
- Conservation International
- Conservation Law Foundation
- Defenders of Wildlife
- Earthwatch
- Earth First!
- Earth Island Institute
- Earth Policy Institute
- Earth Liberation Army (ELA)
- Earth Liberation Front (ELF)
- EarthLab
- Earth's Birthday Project
- Ecotrust
- EcoWatch
- Energy Action Coalition
- Environmental and Energy Study Institute (EESI)
- Environment America
- Environment California
- Environmental Defence Fund
- Environmental Design Research Association (EDRA)
- Environmental Law Institute

- Environmental Life Force (ELF)
- Environmental Working Group
- Earth Share
- Forest Guardians
- Global Water Policy Project
- Green Zionist Alliance
- GREENGUARD Environmental Institute
- Hudson River Sloop Clearwater
- Institute for Energy and Environmental Research (IEER)
- Institute of Environmental Sciences and Technology
- International Council on Nanotechnology (ICON)
- Honour the Earth
- Izaak Walton League
- Keep America Beautiful
- League of Conservation Voters
- Montana Wilderness Association
- National Audubon Society
- National Council for Science and the Environment (NCSE)
- National Geographic Society
- National Registry of Environmental Professionals (NREP)
- National Wildlife Federation
- National Wildlife Refuge Association
- Native Forest Council
- Natural Resources Defence Council
- Nature's Classroom
- NatureServe
- Negative Population Growth
- Neighbourhood Parks Council
- Nicodemus Wilderness Project
- Pacific Environment
- Public Employees for Environmental Responsibility (PEER)
- Population Connection
- Preserve Our Island
- Rainforest Action Network

- Resources for the Future (RFF)
- Republicans for Environmental Protection
- Rising Tide North America
- Riverkeeper
- Sand County Foundation
- Save the Redwoods League
- Science & Environmental Policy Project (SEPP)
- Sea Shepherd
- Sierra Club
- Silicon Valley Toxics Coalition
- Student Conservation Association
- Student Environmental Action Coalition (SEAC)
- Surfrider Foundation
- Sustainable Silicon Valley (SSV)
- Tellus Institute
- Texas Campaign for the Environment
- The Big Green Bus
- The Conservation Fund
- The Marine Mammal Centre
- The Nature Conservancy
- The Ocean Conservancy
- The School for Field Studies
- The Wilderness Society
- TreePeople
- Union of Concerned Scientists
- Waterkeeper Alliance
- West Harlem Environmental Action (WEACT)
- WILD Foundation
- Worldwatch Institute
- Wyoming Outdoor Council
- youthNoise.

Natural Environment

The natural environment encompasses all living and non-living things occurring naturally on Earth or some region thereof. It is an environment that encompasses the interaction of all living species. The concept of the *natural environment* can be distinguished by components:

- Complete ecological units that function as natural systems without massive human intervention, including all vegetation, microorganisms, soil, rocks, atmosphere and natural phenomena that occur within their boundaries.

- Universal natural resources and physical phenomena that lack clear-cut boundaries, such as air, water, and climate, as well as energy, radiation, electric charge, and magnetism, not originating from human activity.

The natural environment is contrasted with the built environment, which comprises the areas and components that are strongly influenced by humans. A geographical area is regarded as a natural environment.

Composition

Earth science generally recognises 4 spheres, the lithosphere, the hydrosphere, the atmosphere, and the biosphere as correspondent to rocks, water, air, and life. Some scientists include, as part of the spheres of the Earth, the cryosphere (corresponding to ice) as a distinct portion of the hydrosphere, as well as the pedosphere (corresponding to soil) as an active and intermixed sphere. Earth science (also known as geoscience, the geosciences or the Earth Sciences), is an all-embracing term for the sciences related to the planet Earth. There are four major disciplines in earth sciences, namely geography, geology, geophysics and geodesy. These major disciplines use physics, chemistry, biology, chronology and mathematics to build a qualitative and quantitative understanding of the principal areas or *spheres* of the Earth system.

Geological Activity

The Earth's crust, or lithosphere, is the outermost solid surface of the planet and is chemically and mechanically different from underlying mantle. It has been generated largely by igneous processes in which magma (molten rock) cools and solidifies to form solid rock. Beneath the lithosphere lies the mantle which is heated by the decay of radioactive elements.

The mantle though solid is in a state of rheic convection. This convection process causes the lithospheric plates to move, albeit slowly. The resulting process is known as plate tectonics. Volcanoes result primarily from the melting of subducted crust material or of rising mantle at mid-ocean ridges and mantle plumes.

Water on Earth

Oceans

An ocean is a major body of saline water, and a component of the hydrosphere. Approximately 71% of the Earth's surface (an area of some 362 million square kilometres) is covered by ocean, a continuous body of water that is customarily divided into several principal oceans and smaller seas. More than half of this area is over 3,000 metres (9,800 ft) deep. Average oceanic salinity is around 35 parts per thousand (ppt) (3.5%), and nearly all seawater has a salinity in the range of 30 to 38 ppt. Though generally recognised as several 'separate' oceans, these waters comprise one global, interconnected body of salt water often referred to as the World Ocean or global ocean. This concept of a global ocean as a continuous body of water with relatively free interchange among its parts is of fundamental importance to oceanography. The major oceanic divisions are defined in part by the continents, various archipelagos, and other criteria: these divisions are (in descending order of size) the Pacific Ocean, the Atlantic Ocean, the Indian Ocean, the Southern Ocean and the Arctic Ocean.

Rivers

Figure : The Columbia River, along the border of the U.S. states of Oregon and Washington.

A river is a natural watercourse, usually freshwater, flowing toward an ocean, a lake, a sea or another river. In a few cases, a river simply flows into the ground or dries up completely before reaching another body of water. Small rivers may also be termed by several other names, including stream, creek and brook. In the United States a river is generally classified as a watercourse more than 60 feet (18 metres) wide. The water in a river is usually in a channel, made up of a stream bed between banks. In larger rivers there is also a wider floodplain shaped by flood-waters over-topping the channel. Flood plains may be very wide in relation to the size of the river channel. Rivers are a part of the hydrological cycle. Water within a river is generally collected from precipitation through surface runoff, groundwater recharge, springs, and the release of water stored in glaciers and snowpacks.

Streams

A stream is a flowing body of water with a current, confined within a bed and stream banks. Streams play an important corridor role in connecting fragmented habitats and thus in conserving biodiversity. The study of streams and waterways in general is known as *surface hydrology*. Types of streams include creeks, tributaries, which do not reach an ocean and connect with another stream or river, brooks, which are typically small streams and sometimes sourced from a spring or seep and tidal inlets.

Lakes

The Lácar Lake is a lake of glacial origin in theprovince of Neuquén, Argentina.

A lake (from Latin *lacus*) is a terrain feature, a body of water that is localised to the bottom of basin. A body of water is considered a lake when it is inland, is not part of a ocean, is larger and deeper than a pond, and is fed by a river.

Natural lakes on Earth are generally found in mountainous areas, rift zones, and areas with ongoing or recent glaciation. Other lakes are found in endorheic basins or along the courses of mature rivers. In some parts of the world, there are many lakes because of chaotic drainage patterns left over from the last Ice Age. All lakes are temporary over geologic time scales, as they will slowly fill in with sediments or spill out of the basin containing them.

A swamp area in Everglades National Park, Florida, USA.

Ponds

A pond is a body of standing water, either natural or man-made, that is usually smaller than a lake. A wide variety of man-made bodies of water are classified as ponds, including water gardens designed for aesthetic ornamentation, fish ponds designed for commercial fish breeding, and solar ponds designed to store thermal energy. Ponds and lakes are distinguished from streams viacurrent speed. While currents in streams are easily observed, ponds and lakes possess thermally driven micro-currents and moderate wind driven currents. These features distinguish a pond from many other aquatic terrain features, such as stream pools and tide pools.

Atmosphere, Climate and Weather

Atmospheric gases scatter blue light more than other wavelengths, creating a blue halo when seen from space.

The atmosphere of the Earth serves as a key factor in sustaining the planetary ecosystem. The thin layer of gases that envelops the Earth is held in place by the planet's gravity. Dry air consists of 78% nitrogen, 21% oxygen, 1% argon and other inert gases, such as carbon dioxide. The remaining gases are often referred to as trace gases, among which are the greenhouse gases such as water vapour, carbon dioxide, methane, nitrous oxide, and ozone.

Filtered air includes trace amounts of many other chemical compounds. Air also contains a variable amount of water vapour and suspensions of water droplets and ice crystals seen as clouds. Many natural substances may be present in tiny amounts in an unfiltered air sample, including dust, pollen and spores, sea spray, volcanic ash, and meteoroids. Various industrial pollutants also may be present, such as chlorine (elementary or in compounds), fluorine compounds, elemental mercury, and sulphur compounds such as sulphur dioxide [SO_2]. The ozone layer of the Earth's atmosphere plays an important role in depleting the amount of ultraviolet (UV) radiation that reaches the surface.

As DNA is readily damaged by UV light, this serves to protect life at the surface. The atmosphere also retains heat during the night, thereby reducing the daily temperature extremes.

Atmospheric Layers

Principal Layers

Earth's atmosphere can be divided into five main layers. These layers are mainly determined by whether temperature increases or decreases with altitude. From highest to lowest, these layers are:

- Exosphere: The outermost layer of Earth's atmosphere extends from the exobase upward, mainly composed of hydrogen and helium.
- Thermosphere: The top of the thermosphere is the bottom of the exosphere, called the exobase. Its height varies with solar activity and ranges from about 350–800 km (220–500 mi; 1,100,000–2,600,000 ft). The International Space Station orbits in this layer, between 320 and 380 km (200 and 240 mi).
- Mesosphere: The mesosphere extends from the stratopause to 80–85 km (50–53 mi; 260,000–280,000 ft). It is the layer where most meteors burn up upon entering the atmosphere.
- Stratosphere: The stratosphere extends from the tropopause to about 51 km (32 mi; 170,000 ft). The stratopause, which is

the boundary between the stratosphere and mesosphere, typically is at 50 to 55 km (31 to 34 mi; 160,000 to 180,000 ft).

- Troposphere: The troposphere begins at the surface and extends to between 7 km (23,000 ft) at the poles and 17 km (56,000 ft) at the equator, with some variation due to weather. The troposphere is mostly heated by transfer of energy from the surface, so on average the lowest part of the troposphere is warmest and temperature decreases with altitude. The tropopause is the boundary between the troposphere and stratosphere.

Other Layers

Within the five principal layers determined by temperature are several layers determined by other properties.

- The ozone layer is contained within the stratosphere. It is mainly located in the lower portion of the stratosphere from about 15–35 km (9.3–22 mi; 49,000–110,000 ft), though the thickness varies seasonally and geographically. About 90% of the ozone in our atmosphere is contained in the stratosphere.

- The ionosphere, the part of the atmosphere that is ionized by solar radiation, stretches from 50 to 1,000 km (31 to 620 mi; 160,000 to 3,300,000 ft) and typically overlaps both the exosphere and the thermosphere. It forms the inner edge of the magnetosphere.

- The homosphere and heterosphere: The homosphere includes the troposphere, stratosphere, and mesosphere. The upper part of the heterosphere is composed almost completely of hydrogen, the lightest element.

- The planetary boundary layer is the part of the troposphere that is nearest the Earth's surface and is directly affected by it, mainly through turbulent diffusion.

Effects of Global Warming

The potential dangers of global warming are being increasingly studied by a wide global consortium of scientists. These scientists are increasingly concerned about the potential long-term effects of global warming on our natural environment and on the planet. Of particular concern is how climate change and global warming caused by anthropogenic, or human-made releases of greenhouse gases, most notably carbon dioxide, can act interactively, and have adverse effects upon the planet, its natural environment and humans' existence.

The Retreat of glaciers since 1850 of Aletsch Glacier in the Swiss Alps (situation in 1979, 1991 and 2002), due to global warming.

Efforts have been increasingly focused on the mitigation of greenhouse gases that are causing climatic changes, on developing adaptative strategies to global warming, to assist humans, animal and plant species, ecosystems, regions and nations in adjusting to the effects of global warming. Some examples of recent collaboration to address climate change and global warming include:

Another view of the Aletsch Glacier in the Swiss Alps and because of global warming it has been decreasing

- The United Nations Framework Convention Treaty and convention on Climate Change, to stabilise greenhouse gas concentrations in the atmosphere at a level that would prevent dangerous anthropogenic interference with the climate system.

- The Kyoto Protocol, which is the protocol to the international Framework Convention on Climate Change treaty, again with the objective of reducing greenhouse gases in an effort to prevent anthropogenic climate change.
- The Western Climate Initiative, to identify, evaluate, and implement collective and cooperative ways to reduce greenhouse gases in the region, focusing on a market-based cap-and-trade system.

A significantly profound challenge is to identify the natural environmental dynamics in contrast to environmental changes not within natural variances. A common solution is to adapt a static view neglecting natural variances to exist. Methodologically, this view could be defended when looking at processes which change slowly and short time series, while the problem arrives when fast processes turns essential in the object of the study.

Climate

Climate encompasses the statistics of temperature, humidity, atmospheric pressure, wind, rainfall, atmospheric particle count and numerous other meteorological elements in a given region over long periods of time. Climate can be contrasted to weather, which is the present condition of these same elements over periods up to two weeks.

Climates can be classified according to the average and typical ranges of different variables, most commonly temperature and precipitation. The most commonly used classification scheme is the one originally developed by Wladimir Köppen. The Thornthwaite system, in use since 1948, incorporates evapotranspiration in addition to temperature and precipitation information and is used in studying animal species diversity and potential impacts of climate changes.

Weather

Weather is a set of all the phenomena occurring in a given atmospheric area at a given time. Most weather phenomena occur in the troposphere, just below the stratosphere. Weather refers, generally, to day-to-day temperature and precipitation activity, whereas climate is the term for the average atmospheric conditions over longer periods of time. When used without qualification, "weather" is understood to be the weather of Earth.

Weather occurs due to density (temperature and moisture) differences between one place and another. These differences can occur due to the sun angle at any particular spot, which varies by latitude from the tropics. The strong temperature contrast between polar and tropical air gives rise to the jet stream. Weather systems in the mid-latitudes, such as extratropical cyclones, are caused by instabilities of the jet stream flow. Because the Earth's axis is tilted relative to its orbital plane, sunlight is incident at different angles at different times of the year. On the Earth's surface, temperatures usually range ±40 °C (100 °F to "40 °F) annually. Over thousands of years, changes in the Earth's orbit have affected the amount and distribution of solar energy received by the Earth and influence long-term climate

Surface temperature differences in turn cause pressure differences. Higher altitudes are cooler than lower altitudes due to differences in compressional heating. Weather forecasting is the application of science and technology to predict the state of the atmosphere for a future time and a given location. The atmosphere is a chaotic system, and small changes to one part of the system can grow to have large effects on the system as a whole. Human attempts to control the weather have occurred throughout human history, and there is evidence that human activity such as agriculture and industry has inadvertently modified weather patterns.

Life

Evidence suggests that life on Earth has existed for about 3.7 billion years. All known life forms share fundamental molecular mechanisms, and based on these observations, theories on the origin of life attempt to find a mechanism explaining the formation of a primordial single cell organism from which all life originates. There are many different hypotheses regarding the path that might have been taken from simple organic molecules via precellular life to protocells and metabolism.

Although there is no universal agreement on the definition of life, scientists generally accept that the biological manifestation of life is characterised by organisation, metabolism, growth, adaptation, response to stimuli and reproduction. Life may also be said to be simply the characteristic state of organisms. In biology, the science of living organisms, "life" is the condition which distinguishes active organisms from inorganic matter, including the capacity for growth, functional activity and the continual change preceding death.

A diverse array of living organisms (life forms) can be found in the biosphere on Earth, and properties common to these organisms—plants, animals, fungi, protists, archaea, and bacteria—are a carbon- and water-based cellular form with complex organisation and heritable genetic information. Living organisms undergo metabolism, maintain homeostasis, possess a capacity to grow, respond to stimuli, reproduce and, through natural selection, adapt to their environment in successive generations. More complex living organisms can communicate through various means.

Ecosystems

Rainforests often have a great deal of biodiversity with many plant and animal species. This is the Gambia River in Senegal'sNiokolo-Koba National Park.

An ecosystem (also called as environment) is a natural unit consisting of all plants, animals and micro-organisms (biotic factors) in an area functioning together with all of the non-living physical (abiotic) factors of the environment.

Central to the ecosystem concept is the idea that living organisms are continually engaged in a highly interrelated set of relationships with every other element constituting the environment in which they exist. Eugene Odum, one of the founders of the science of ecology, stated: "Any unit that includes all of the organisms (ie: the "community") in a given area interacting with the physical environment so that a flow of energy leads to clearly defined trophic structure, biotic diversity, and material cycles (i.e.: exchange of materials between living and nonliving parts) within the system is an ecosystem."

The human ecosystem concept is then grounded in the deconstruction of the human/nature dichotomy, and the emergent premise that all species are ecologically integrated with each other, as well as with the abiotic constituents of their biotope.

Old-growth forest and a creek on Larch Mountain, in the U.S. state of Oregon.

A greater number or variety of species or biological diversity of an ecosystem may contribute to greater resilience of an ecosystem, because there are more species present at a location to respond to change and thus "absorb" or reduce its effects. This reduces the effect before the ecosystem's structure is fundamentally changed to a different state. This is not universally the case and there is no proven relationship between the species diversity of an ecosystem and its ability to provide goods and services on a sustainable level.

The term ecosystem can also pertain to human-made environments, such as human ecosystems and human-influenced ecosystems, and can describe any situation where there is relationship between living organisms and their environment. Fewer areas on the surface of the earth today exist free from human contact, although some genuine wilderness areas continue to exist without any forms of human intervention.

Biomes

Biomes are terminologically similar to the concept of ecosystems, and are climatically and geographically defined areas of ecologically similar climatic conditions on the Earth, such as communities of plants, animals, and soil organisms, often referred to *as* ecosystems. Biomes are defined on the basis of factors such as plant structures (such as trees, shrubs, and grasses), leaf types (such as broadleaf and needle leaf), plant spacing (forest, woodland, savanna), and climate. Unlike ecozones, biomes are not defined by genetic, taxonomic, or historical similarities. Biomes are often identified with particular patterns of ecological succession and climax vegetation.

Biogeochemical Cycles

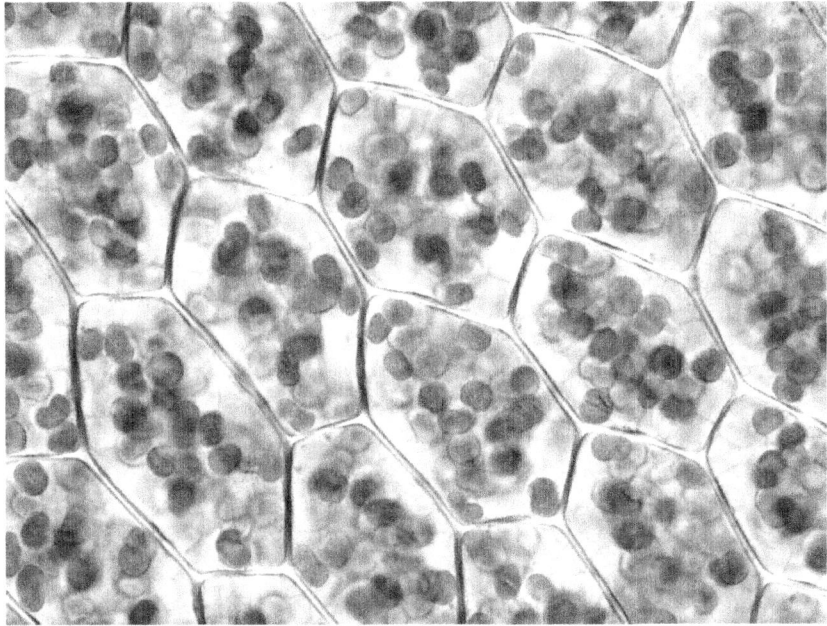

Chloroplasts conduct photosynthesis and are found in plant cells and other eukaryoticorganisms. These are Chloroplasts visible in the cells of Plagiomnium affine - Many-fruited Thyme-moss.: Biogeochemical cycles

Global biogeochemical cycles are critical to life, most notably those of water, oxygen, carbon, nitrogen and phosphorus.

- The nitrogen cycle is the transformation of nitrogen and nitrogen-containing compounds in nature. It is a cycle which includes gaseous components.

- The water cycle, is the continuous movement of water on, above, and below the surface of the Earth. Water can change states among liquid, vapour, and ice at various places in the water cycle. Although the balance of water on Earth remains fairly constant over time, individual water molecules can come and go.

- The carbon cycle is the biogeochemical cycle by which carbon is exchanged among the biosphere, pedosphere, geosphere, hydrosphere, and atmosphere of the Earth.

- The oxygen cycle is the movement of oxygen within and between its three main reservoirs: the atmosphere, the biosphere, and the lithosphere. The main driving factor of the oxygen cycle is photosynthesis, which is responsible for the modern Earth's atmospheric composition and life.

- The phosphorus cycle is the movement of phosphorus through the lithosphere, hydrosphere, and biosphere. The atmosphere does not play a significant role in the movements of phosphorus, because phosphorus and phosphorus compounds are usually solids at the typical ranges of temperature and pressure found on Earth.

Wilderness

A conifer forest in the Swiss Alps (National Park).

Wilderness is generally defined as a natural environment on Earth that has not been significantly modified by human activity. The WILD Foundation goes into more detail, defining wilderness as: "The most intact, undisturbed wild natural areas left on our planet- those last truly wild places that humans do not control and have not developed with roads, pipelines or other industrial infrastructure." Wilderness areas and protected parks are considered important for the survival of certain species, ecological studies, conservation, solitude, and recreation. Wilderness is deeply valued for cultural, spiritual, moral, and aesthetic reasons. Some nature writers believe wilderness areas are vital for the human spirit and creativity.

The word, "wilderness", derives from the notion of wildness; in other words that which is not controllable by humans. The word'setymology is from the Old English *wildeornes*, which in turn derives from *wildeor* meaning *wild beast* (wild + deor = beast, deer). From this point of view, it is the wildness of a place that makes it a wilderness. The mere presence or activity of people does not disqualify an area from being "wilderness." Many ecosystems that are, or have been, inhabited or influenced by activities of people may still be considered "wild." This way of looking at wilderness includes areas within which natural processes operate without very noticeable human interference.

Wildlife includes all non-domesticated plants, animals and other organisms. Domesticating wild plant and animal species for human benefit has occurred many times all over the planet, and has a major impact on the environment, both positive and negative. Wildlife can be found in all ecosystems. Deserts, rain forests, plains, and other areas—including the most developed urban sites—all have distinct forms of wildlife. While the term in popular culture usually refers to animals that are untouched by human factors, most scientists agree that wildlife around the world is impacted by human activities.

Challenges

Before flue gas desulfurization was installed, the air-polluting emissions from this power plant in New Mexicocontained excessive amounts of sulfur dioxide

Amazon Rainforest in Brazil. The tropical rainforests of South America contain the largest diversity of species on Earth, including some that have evolved within the past few hundred thousand years.

It is the common understanding of *natural environment* that underlies environmentalism — a broad political, social, and philosophical movement that advocates various actions and policies in the interest of protecting what nature remains in the natural environment, or restoring or expanding the role of nature in this environment. While true wilderness is increasingly rare, *wild* nature (e.g., unmanaged forests, uncultivated grasslands, wildlife, wildflowers) can be found in many locations previously inhabited by humans.

Goals commonly expressed by environmental scientists include:

- Reduction and clean up of pollution, with future goals of zero pollution;
- Cleanly converting non-recyclable materials into energy through direct combustion or after conversion into secondary fuels;
- Reducing societal consumption of non-renewable fuels;
- Development of alternative, green, low-carbon or renewable energy sources;
- Conservation and sustainable use of scarce resources such as water, land, and air;
- Protection of representative or unique or pristine ecosystems;
- Preservation of threatened and endangered species extinction;
- The establishment of nature and biosphere reserves under various types of protection; and, most generally, the protection of biodiversity and ecosystems upon which all human and other life on earth depends.

Very large development projects- megaprojects- pose special instructions and risks to the natural environments. Major dams and power plants are cases in point. The challenge to the environment from such projects is growing because more and bigger megaprojects are being built, in developed and developing nations alike.

Cultural Ecology

Cultural ecology studies the relationship between a given society and its natural environment as well as the life-forms and ecosystems that support its lifeways. This may be carried out diachronically (examining entities that existed in different epochs), or synchronically (examining a present system and its components). The central argument is that the natural environment, in small scale or subsistence societies dependent in part upon it- is a major contributor to social organisation and other human institutions.

In the academic realm, when combined with study of political economy, the study of economies as polities, it becomes political ecology, another academic subfield. It also helps interrogate historical events like the Easter Island Syndrome.

Coining the Term

Anthropologist Julian Steward (1902-1972) coined the term, envisioning cultural ecology as a methodology for understanding how humans adapt to such a wide variety of environments. In his *Theory of Culture Change: The Methodology of Multilinear Evolution* (1955), cultural ecology represents the "ways in which culture change is induced by adaptation to the environment." A key point is that any particular human adaptation is in part historically inherited and involves the technologies, practices, and knowledge that allow people to live in an environment. This means that while the environment influences the character of human adaptation, it does not determine it. In this way, Steward wisely separated the vagaries of the environment from the inner workings of a culture that occupied a given environment. Viewed over the long term, this means that environment and culture are on more or less separate evolutionary tracks and that the ability of one to influence the other is dependent on how each is structured.

It is this assertion- that the physical and biological environment affects culture- that has proved controversial, because it implies an element of environmental determinism over human actions, which some social scientists find problematic, particularly those writing from a Marxist perspective. Cultural ecology recognises that ecological locale plays a significant role in shaping the cultures of a region. Steward's method was to:

1. document the technologies & methods used to exploit the environment- to get a living from it.

2. look at patterns of human behaviour/culture associated with using the environment.

3. assess how much these patterns of behaviour influenced other aspects of culture (e.g., how, in a drought-prone region, great concern over rainfall patterns meant this became central to everyday life, and led to the development of a religious belief system in which rainfall and water figured very strongly. This belief system may not appear in a society where good rainfall for crops can be taken for granted, or where irrigation was practiced).

Steward's concept of cultural ecology became widespread among anthropologists and archaeologists of the mid-20th century, though they would later be critiqued for their environmental determinism. Cultural ecology was one of the central tenets and driving factors in the development of processual archaeology in the 1960s, as archaeologists understood cultural change through the framework of technology and its effects on environmental adaptation.

Cultural Ecology in Anthropology

Cultural ecology as developed by Steward is a major subdiscipline of anthropology. It derives from the work of Franz Boas and has branched out to cover a number of aspects of human society, in particular the distribution of wealth and power in a society, and how that affects such behaviour as hoarding or gifting (e.g. the Haida tradition of the potlatch on the Canadian west-coast).

Cultural Ecology as a Transdisciplinary Project

One recent conception of cultural ecology is as a general theory that regards ecology as a paradigm not only for the natural and human sciences, but for cultural studies as well. In his *Die Ökologie des Wissens* (The Ecology of Knowledge), Peter Finke explains that this theory brings together the various cultures of knowledge that have evolved in history, and that have been separated into more and more specialised disciplines and subdisciplines in the evolution of modern science (Finke 2005). In this view, cultural ecology considers the sphere of human culture not as separate from but as interdependent with and transfused by ecological processes and natural energy cycles. At the same time, it recognises the relative independence and self-reflexive dynamics of cultural processes. As the dependency of culture on nature, and the ineradicable presence of nature in culture, are gaining interdisciplinary attention, the difference between cultural evolution and natural evolution is increasingly acknowledged by cultural ecologists. Rather than genetic laws, information and communication have become major driving forces of cultural evolution. Thus, causal deterministic laws do not apply to culture in a strict sense, but there are nevertheless productive analogies that can be drawn between ecological and cultural processes.

Gregory Bateson was the first to draw such analogies in his project of an Ecology of Mind (Bateson 1973), which was based on general principles of complex dynamic life processes, e.g. the concept of feedback loops, which he saw as operating both between the mind and the world and within the mind itself. Bateson thinks of the mind neither as an autonomous metaphysical force nor as a mere neurological function of the brain, but as a "dehierarchized concept of a mutual dependency between the (human) organism and its (natural) environment, subject and object, culture and nature", and thus as "a synonym for a cybernetic system of information circuits that are relevant for the survival of the species."

Finke fuses these ideas with concepts from systems theory. He describes the various sections and subsystems of society as 'cultural ecosystems' with their own processes of production, consumption, and reduction of energy (physical as well as psychic energy). This also applies to the cultural ecosystems of art and of literature, which follow their own internal forces of selection and self-renewal, but also have an important function within the cultural system as a whole.

Cultural Ecology in Literary Studies

The vital interrelatedness between culture and nature has been a special focus of literary culture from its archaic beginnings in myth, ritual, and oral story-telling, in legends and fairy tales, in the genres of pastoral literature, nature poetry. Important texts in this tradition include the stories of mutual transformations between human and nonhuman life, most famously collected in Ovid's Metamorphoses, which became a highly influential text throughout literary history and across different cultures. This attention to culture-nature interaction became especially prominent in the era of romanticism, but continues to be characteristic of literary stagings of human experience up to the present. The mutual opening and symbolic reconnection of culture and nature, mind and body, human and nonhuman life in a holistic and yet radically pluralistic way seems to be one significant mode in which literature functions and in which literary knowledge is produced. From this perspective, literature can itself be described as the symbolic medium of a particularly powerful form of "cultural ecology" (Zapf 2002). Literary texts have staged and explored, in ever new scenarios, the complex feedback relationship of prevailing cultural systems with the needs and manifestations of human and nonhuman "nature." From this paradoxical act of creative regression they have derived their specific power of innovation and cultural self-renewal.

German ecocritic Hubert Zapf argues that literature draws its cognitive and creative potential from a threefold dynamics in its relationship to the larger cultural system: as a "cultural-critical metadiscourse," an "imaginative counterdiscourse," and a "reintegrative interdiscourse" (Zapf 2001, 2002). It is a textual form which breaks up ossified social structures and ideologies, symbolically empowers the marginalised, and reconnects what is culturally separated. In that way, literature counteracts economic, political or pragmatic forms of interpreting and instrumentalizing human life, and breaks up one-dimensional views of the world and the self, opening them up towards their repressed or excluded other. Literature is thus, on the one hand, a sensorium for what goes wrong in a society, for the biophobic, life-paralysing implications of one-sided forms of consciousness and civilizational uniformity, and it is, on the other hand, a medium of constant cultural self-renewal, in which the neglected biophilic energies can find a symbolic space of expression and of (re-)integration into the larger ecology of cultural discourses. This approach has been applied and widened in a recent volume of essays by scholars from over the world (Zapf 2008).

Cultural Ecology in Geography

In geography, cultural ecology developed in response to the "landscape morphology" approach of Carl O. Sauer. Sauer's school was criticized for being unscientific and holding an [what? missing text] of cultural ecology applied ideas from ecology and systems theory to understand the adaptation of humans to their environment. These cultural ecologists focused on flows of energy and materials, examining how beliefs and institutions in a culture regulated its interchanges with the natural ecology that surrounded it. In this perspective humans were as much a part of the ecology as any other organism. Important practitioners of this form of cultural ecology include Karl Butzer and David Stoddard.

The second form of cultural ecology introduced decision theory from agricultural economics, particularly inspired by the works of

Alexander Chayanov and Ester Boserup. These cultural ecologists were concerned with how human groups made decisions about how they use their natural environment. They were particularly concerned with the question of agricultural intensification, refining the competing models of Thomas Malthus and Boserup. Notable cultural ecologists in this second tradition include Harold Brookfield and Billie Lee Turner II.

Starting in the 1980s, cultural ecology came under criticism from political ecology. Political ecologists charged that cultural ecology ignored the connections between the local-scale systems they studied and the global political economy. Today few geographers self-identify as cultural ecologists, but ideas from cultural ecology have been adopted and built on by political ecology, land change science, and sustainability science.

Conceptual Views of Culture and Ecology

The Human Species

Books about culture and ecology began to emerge in the 1950s and 1960s. One of the first to be published in the United Kingdom was *The Human Species* by a zoologist, Anthony Barnett. It came out in 1950-subtitled *The biology of man* but was about a much narrower subset of topics. It dealt with the cultural bearing of some outstanding areas of environmental knowledge about health and disease, food, the sizes and quality of human populations, and the diversity of human types and their abilities. Barnett's view was that his selected areas of information "....are all topics on which knowledge is not only desirable, but for a twentieth-century adult, necessary". He went on to point out some of the concepts underpinning human ecology towards the social problems facing his readers in the 1950s as well as the assertion that human nature cannot change, what this statement could mean, and whether it is true. The third chapter deals in more detail with some aspects of human genetics.

Then come five chapters on the evolution of man, and the differences between groups of men (or races) and between individual men and women today in relation to population growth (the topic of 'human diversity'). Finally, there is a series of chapters on various aspects of human populations (the topic of "life and death"). Like other animals man must, in order to survive, overcome the dangers of starvation and infection; at the same time he must be fertile. Four chapters therefore deal with food, disease and the growth and decline of human populations.

Barnett anticipated that his personal scheme might be criticised on the grounds that it omits an account of those human characteristics, which distinguish humankind most clearly, and sharply from other animals. That is to say, the point might be expressed by saying that human behaviour is ignored; or some might say that human psychology is left out, or that no account is taken of the human mind. He justified his limited view, not because little importance was attached to what was left out, but because the omitted topics were so important that each needed a book of similar size even for a summary account. In other words, the author was embedded in a world of academic specialists and therefore somewhat worried about taking a partial conceptual, and idiosyncratic view of the zoology of *Homo sapiens*.

The Ecology of Man

Moves to produce prescriptions for adjusting human culture to ecological realities were also afoot in North America. Paul Sears, in his 1957 Condon Lecture at the University of Oregon, titled "The Ecology of Man," he mandated "serious attention to the ecology of man" and demanded "its skillful application to human affairs." Sears was one of the few prominent ecologists to successfully write for popular audiences. Sears documents the mistakes American farmers made in creating conditions that led to the disastrous Dust Bowl. This book gave momentum to the soil conservation movement in the United States.

Man's Impact on Nature

During this same time was J.A. Lauwery's *Man's Impact on Nature*, which was part of a series on 'Interdependence in Nature' published in 1969. Both Russel's and Lauwerys' books were about cultural ecology, although not titled as such. People still had difficulty in escaping from their labels. Even *Beginnings and Blunders*, produced in 1970 by the polymath zoologist Lancelot Hogben, with the subtitle *Before Science Began*, clung to anthropology as a traditional reference point. However, its slant makes it clear that 'cultural ecology' would be a more apt title to cover his wide-ranging description of how early societies adapted to environment with tools, technologies and social groupings. In 1973 the physicist Jacob Bronowski produced *The Ascent of Man*, which summarised a magnificent thirteen part BBC television series about all the ways in which humans have moulded the Earth and its future.

Changing the Face of the Earth

By the 1980s the human ecological-functional view had prevailed. It had become a conventional way to present scientific concepts in the ecological perspective of human animals dominating an overpopulated world, with the practical aim of producing a greener culture. This is exemplified by I. G. Simmons book *Changing the Face of the Earth*, with its telling subtitle "Culture, Environment History" which was published in 1989. Simmons was a geographer, and his book was a tribute to the influence of W.L Thomas' edited collection, *Man's role in 'Changing the Face of the Earth* that came out in 1956.

Simmons' book was one of many interdisciplinary culture/environment publications of the 1970s and 1980s, which triggered a crisis in geography with regards its subject matter, academic subdivisions, and boundaries. This was resolved by officially adopting conceptual frameworks as an approach to facilitate the organisation of research and teaching that cuts cross old subject divisions. Cultural ecology is in fact a conceptual arena that has, over the past six decades allowed sociologists, physicists, zoologists and geographers to enter common intellectual ground from the sidelines of their specialist subjects.

Relationship in the 21st Century

In the first decade of the 21st century, there are publications dealing with the ways in which humans can develop a more acceptable cultural relationship with the environment. An example is sacred ecology, a sub-topic of cultural ecology, produced by Fikret Berkes in 1999. It seeks lessons from traditional ways of life in Northern Canada to shape a new environmental perception for urban dwellers. This particular conceptualisation of people and environment comes from various cultural levels of local knowledge about species and place, resource management systems using local experience, social institutions with their rules and codes of behaviour, and a world view through religion, ethics and broadly defined belief systems.

Despite the differences in information concepts, all of the publications carry the message that culture is a balancing act between the mindset devoted to the exploitation of natural resources and that, which conserves them. Perhaps the best model of cultural ecology in this context is, paradoxically, the mismatch of culture and ecology that have occurred when Europeans suppressed the age-old native methods of land use and have tried to settle European farming cultures on soils manifestly incapable of supporting them. There is a sacred ecology associated with environmental awareness, and the task of cultural ecology is to inspire urban dwellers to develop a more acceptable sustainable cultural relationship with the environment that supports them.

Ecohydrology

Ecohydrology is an interdisciplinary field studying the interactions between water and ecosystems. These interactions may take place within water bodies, such as rivers and lakes, or on land, in forests, deserts, and other terrestrial ecosystems. Areas of research in ecohydrology include transpiration and plant water use, adaption of organisms to their water environment, influence of vegetation on stream flow and function, and feedbacks between ecological processes and the hydrological cycle.

Key concepts

The hydrologic cycle describes the continuous movement of water on, above, and below the surface on the earth. This flow is altered by ecosystems at numerous points. Transpiration from plants provides the majority of flow of water to the atmosphere. Water is influenced by vegetative cover as it flows over the land surface, while river channels can be shaped by the vegetation within them. Ecohydrologists study both terrestrial and aquatic systems. In terrestrial ecosystems (such as forests, deserts, and savannas), the interactions among vegetation, the land surface, the vadose zone, and the groundwater are the main focus. In aquatic ecosystems (such as rivers, streams, lakes, and wetlands), emphasis is placed on how water chemistry, geomorphology, and hydrology affect their structure and function.

Principles

The principles of Ecohydrology are expressed in three sequential components:

1. Hydrological: The quantification of the hydrological cycle of a basin, should be a template for functional integration of hydrological and biological processes.

2. Ecological: The integrated processes at river basin scale can be steered in such a way as to enhance the basin's carrying capacity and its ecosystem services.

3. Ecological engineering: The regulation of hydrological and ecological processes, based on an integrative system approach, is thus a new tool for Integrated Water Basin Management.

Their expression as testable hypotheses (Zalewski et al., 1997) may be seen as:

• H1: Hydrological processes generally regulate biota

• H2: Biota can be shaped as a tool to regulate hydrological processes

- H3: These two types of regulations (H1&H2) can be integrated with hydro-technical infrastructure to achieve sustainable water and ecosystem services.

Vegetation and Water Stress

A fundamental concept in ecohydrology is that plant physiology is directly linked to water availability. Where there is ample water, as in rainforests, plant growth is more dependent on nutrient availability. However, in semi-arid areas, like African savannas, vegetation type and distribution relate directly to the amount of water that plants can extract from the soil. When insufficient soil water is available, a water-stressed condition occurs. Plants under water stress decrease both their transpiration and photosynthesis through a number of responses, including closing their stomata. This decrease in the canopy water flux and carbon dioxide flux can have an impact on surrounding climate and weather.

Soil Moisture Dynamics

Soil moisture is a general term describing the amount of water present in the vadose zone, or unsaturated portion of soil below ground. Since plants depend on this water to carry out critical biological processes, soil moisture is integral to the study of ecohydrology. Soil moisture is generally described as water content, è, or saturation, S. These terms are related by porosity, n, through the equation $è = nS$. The changes in soil moisture over time are known as soil moisture dynamics.

Temporal and Spatial Considerations

Ecohydrological theory also places importance on considerations of temporal (time) and spatial (space) relationships. Hydrology, in particular the timing of precipitation events, can be a critical factor in the way an ecosystem evolves over time. For instance, Mediterranean landscapes experience dry summers and wet winters. If the vegetation has a summer growing season, it often experiences water stress, even though the total precipitation throughout the year may be moderate. Ecosystems in these regions have typically evolved to support high water demand grasses in the winter, when water availability is high, and drought-adapted trees in the summer, when it is low.

Ecohydrology also concerns itself with the hydrological factors behind the spatial distribution of plants. The optimal spacing and spatial organisation of plants is at least partially determined by water availability. In ecosystems with low soil moisture, trees are typically located further apart than they would be in well-watered areas.

Basic Equations and Models

Water Balance at a Point

A fundamental equation in ecohydrology is the water balance at a point in the landscape. A water balance states that the amount water entering the soil must be equal to the amount of water leaving the soil plus the change in the amount of water stored in the soil. The water balance has four main components: infiltration of precipitation into the soil, evapotranspiration, leakage of water into deeper portions of the soil not accessible to the plant, and runoff from the ground surface. It is described by the following equation:

$$nZ_r \frac{ds(t)}{dt} = R(t) - I(t) - Q[s(t),t] - E[s(t)] - L[s(t)]$$

The terms on the left hand side of the equation describe the total amount of water contained in the rooting zone. This water, accessible to vegetation, has a volume equal to the porosity of the soil (n) multiplied by its saturation (s) and the depth of the plant's roots (Z_r). The differential equation $ds(t)/dt$ describes how the soil saturation changes over time. The terms on the right hand side describe the rates of rainfall (R), interception (I), runoff (Q), evapotranspiration (E), and leakage (L). These are typically given in millimetres per day (mm/d). Runoff, evaporation, and leakage are all highly dependent on the soil saturation at a given time.

In order to solve the equation, the rate of evapotranspiration as a function of soil moisture must be known. The model generally used to describe it states that above a certain saturation, evaporation will only be dependent on climate factors such as available sunlight. Once below this point, soil moisture imposes controls on evapotranspiration, and it decreases until the soil reaches the point where the vegetation can no longer extract any more water. This soil level is generally referred to as the "permanent wilting point". This term is confusing because many plant species do not actually "wilt".

Chapter 6

Ecological Forecasting

Ecological forecasting uses knowledge of physics, ecology and physiology to predict how ecosystems will change in the future in response to environmental factors such as climate change. The ultimate goal of the approach is to provide people such as resource managers and designers of marine reserves with information that they can then use to respond, in advance, to future changes, a form of adaptation to global warming.

One of the most important environmental factors for organisms today is global warming. Most physiological processes are affected by temperature, and so even small changes in weather and climate can lead to large changes in the growth, reproduction and survival of animals and plants. The scientific consensus is that the increase in atmospheric greenhouse gases due to human activity caused most of the warming observed since the start of the industrial era. These changes are in turn affecting human and natural ecosystems.

One major challenge is to predict where, when and with what magnitude changes are likely to occur so that we can mitigate or at least prepare for them. Ecological forecasting applies existing knowledge of how animals and plants interact with their physical environment to ask how changes in environmental factors might result in changes to the ecosystems as a whole.

Approaches

- Palaeobiology modelling: uses fossil and phylogenetic evidence of biodiversity in the past to project the trajectory of biodiversity in the future. Simple plots can be constructed and then adjusted based on the varying quality of the fossil record.
- Climate envelope modelling: relies on statistical correlations between existing species distributions and environmental variables to define a species' tolerance. *Envelopes* of tolerance are then drawn around existing ranges. By predicting future

levels of factors such as temperature, rainfall, and salinity, new range boundaries are then predicted. These methods are good for examining large numbers of species, but are likely not a good means of predicting effects at fine scales.

- Niche level modelling: is a newer method which links physiological information about a species to models of animal and plant body temperature. In contrast to "climate envelope" approaches, environmental variables are predicted at the level of the niche and are therefore much more exact. However, the approach is also usually more time consuming.

Forecasting Examples

Biodiversity

Using fossil evidence, studies have shown that vertebrate biodiversity has grown exponentially through Earth's history and that biodiversity is entwined with the diversity of Earth's habitats.

"Animals have not yet invaded 2/3 of Earth's habitats,
and it could be that without human influence biodiversity
will continue to increase in an exponential fashion."
—*Sahney* et al.

Temperature

Forecasts of temperature, shown in the diagram at the right as coloured dots, along the North Island of New Zealand in the austral summer of 2007. As per the temperature scale shown at the bottom, intertidal temperatures were forecast to exceed 30°C at some locations on February 19; surveys later showed that these sites corresponded to large die-offs in burrowing sea urchins.

Agroecology

Agroecology is the application of ecological principles to the production of food, fuel, fibre, and pharmaceuticals. The term encompasses a broad range of approaches, and is considered "a science, a movement, and a practice."

Ecological Strategy

Agroecologists study a variety of agroecosystems, and the field of agroecology is not associated with any one particular method of farming, whether it be organic, conventional, intensive or extensive. Furthermore, it is not defined by certain management practices, such as the use of natural enemies in place of insecticides, or polyculture in place of monoculture.

Additionally, agroecologists do not unanimously oppose technology or inputs in agriculture but instead assess how, when, and if technology can be used in conjunction with natural, social and human assets. Agroecology proposes a context- or site-specific manner of studying agroecosystems, and as such, it recognises that there is no universal formula or recipe for the success and maximum well-being of an agroecosystem.

Instead, agroecologists may study questions related to the four system properties of agroecosystems: productivity, stability, sustainability and equitability. As opposed to disciplines that are concerned with only one or some of these properties, agroecologists see all four properties as interconnected and integral to the success of an agroecosystem. Recognising that these properties are found on varying spatial scales, agroecologists do not limit themselves to the study of agroecosystems at any one scale: farm, community, or global.

Agroecologists study these four properties through an interdisciplinary lens, using natural sciences to understand elements of agroecosystems such as soil properties and plant-insect interactions, as well as using social sciences to understand the effects of farming practices on rural communities, economic constraints to developing new production methods, or cultural factors determining farming practices.

Various Approaches

Agroecologists do not always agree about what agroecology is or should be in the long-term. Different definitions of the term agroecology can be distinguished largely by the specificity with which one defines the term "ecology," as well as the term's potential political connotations. Definitions of agroecology, therefore, may be first grouped according to the specific contexts within which they situate agriculture. Agroecology is defined by the OECD as "the study of the relation of agricultural crops and environment." This definition refers to the "-ecology" part of "agroecology" narrowly as the natural environment. Following this definition, an agroecologist would study agriculture's various relationships with soil health, water quality, air quality, meso- and micro-fauna, surrounding flora, environmental toxins, and other environmental contexts.

A more common definition of the word can be taken from Dalgaard et al., who refer to agroecology as the study of the interactions between plants, animals, humans and the environment within agricultural systems. Consequently, agroecology is inherently multidisciplinary, including factors from agronomy, ecology, sociology and economics. In this case, the "-ecology" portion of "agroecology is defined broadly to include social, cultural, and economic contexts as well.

Agroecology is also defined differently according to geographic location. In the global south, the term often carries overtly political connotations. Such political definitions of the term usually ascribe to it the goals of social and economic justice; special attention, in this case, is often paid to the traditional farming knowledge of indigenous populations. North American and European uses of the term sometimes avoid the inclusion of such overtly political goals. In these cases, agroecology is seen more strictly as a scientific discipline with less specific social goals.

Fred Buttel makes a more academic distinction of the various approaches within the field, separating it into five broad categories:

Ecosystems Agroecology

This approach is driven by the ecosystems biology of Eugene Odum. This approach is based in the hypotheses that the natural systems, with its stability and resilience, provide the best model to mimic if sustainability is the goal. Normally, ecosystems agroecology is not actively involved in social science; however, this school is essentially based on the belief that large-scale agriculture is inappropriate. The work of Steve Gliessman is prototypical of this approach.

Agronomic Ecology

The basic approach in this branch is derived mostly from agronomy, including the traditional agricultural production sciences. This approach also does not actively involve social sciences in the agroecological analysis, but uses social sciences to understand the processes by which agriculture became unsustainable. Chuck Francis, Richard Hardwood, Ricardo Salvador, and Matt Liebman are exemplars of this approach.

Ecological Political Economy

The driving force behind this form of agroecology is a political-economical critique of modern agriculture. The school believes that only radical changes in political economy and the moral economy of research will reduce the negative costs of modern agriculture. The works of Miguel Altieri (ecosystem biologist), John Vandermeer (population ecologist), Richard Lewontin, and Richard Levins provide examples of this politically charged and socially-oriented version of agroecology.

Agro-population Ecology

This approach is derived from the science of ecology primarily based on population ecology, which over the past three decades has been displacing the ecosystems biology of Odum. Buttel explains the main difference between the two categories, saying that "the application of population ecology to agroecology involves the primacy not only of analysing agroecosystems from the perspective of the population dynamics of their constituent species, and their relationships to climate and biogeochemistry, but also there is a major emphasis placed on the role of genetics." David Andow and Alison Power are cited as examples of professionals espousing this view.

Integrated Assessment of Multifunctional Agricultural Systems

This approach focuses on the multifunctionality of the landscape, instead of focusing solely on the agricultural enterprise. Agriculture and the food system are considered parts of an institutional complex that relates to and integrates with other social institutions. Scholars adopting this highly integrated approach, mostly Europeans, do not consider any one discipline the leader of agroecology.

Holon Agroecology

First introduced in 2007 by the soil scientist William L. Bland and the environmental sociologist Michael M. Bell of the University of Wisconsin–Madison, holon agroecology draws on Koestler's notion of a "holon" which is both part and whole and develops it with ideas of narrative, intentionality, and incompleteness or unfinalizability, within an ever-changing "ecology of contexts". In contrast to systems thinking, holon agroecology stresses seeing the agricultural endeavour as an unfinished accomplishment that is constantly adjusting itself to its many contexts and their conflicts and incommensurabilities. The farm holon represents a kind of "holding together" in order to persist through change, but a holding together that is never fully unified and worked out.

History

Pre-WWII

The notions and ideas relating to crop ecology have been around since at least 1911 when F.H. King released *Farmers of Forty Centuries*. King was one of the pioneers as a proponent of more quantitative methods for characterisation of water relations and physical properties of soils. In the late 1920s the attempt to merge agronomy and ecology was born with the development of the field of crop ecology. Crop ecology's main concern was where crops would be best grown. Actually, it was only in 1928 that agronomy and ecology were formally linked by Klages.

The first mention of the term agroecology was in 1928, with the publication of the term by Bensin in 1928. The book of Tischler (1965), was probably the first to be actually titled 'agroecology'. He analysed the different components (plants, animals, soils and climate) and their interactions within an agroecosystem as well as the impact of human agricultural management on these components. Other books dealing with agroecology, but without using the term explicitly were published by the German zoologist Friederichs (1930) with his book on agricultural zoology and related ecological/environmental factors for plant protection and by American crop physiologist Hansen in 1939 when both used the word as a synonym for the application of ecology within agriculture.

Post-WWII

Gliessman mentions that post-WWII, groups of scientists with ecologists gave more focus to experiments in the natural environment, while agronomists dedicated their attention to the cultivated systems in agriculture. According to Gliessman, the two groups kept their research and interest apart until books and articles using the concept of agroecosystems and the word agroecology started to appear in 1970. Dalgaard explains the different points of view in ecology schools, and the fundamental differences, which set the basis for the development of agroecology. The early ecology school of Henry Gleason investigated plant populations focusing in the hierarchical levels of the organism under study.

Friederich Clement's ecology school, however included the organism in question as well as the higher hierarchical levels in its investigations, a "landscape perspective". However, the ecological schools where the roots of agroecology lie are even broader in nature. The ecology school of Tansley, whose view included both the biotic organism and their environment, is the one from which the concept of agroecosystems emerged in 1974 with Harper.

In the 1960s and 70's the increasing awareness of how humans manage the landscape and its consequences set the stage for the necessary cross between agronomy and ecology. Even though, in many ways the environmental movement in the US was a product of the times, the Green Decade spread an environmental awareness of the unintended consequences of changing ecological processes. Works such as *Silent Spring*, and *The Limits to Growth*, and changes in legislation such as the Clean Air Act, Clean Water Act, and the

National Environmental Policy Act caused the public to be aware of societal growth patterns, agricultural production, and the overall capacity of the system.

Fusion with Ecology

After the 1970s, when agronomists saw the value of ecology and ecologists began to use the agricultural systems as study plots, studies in agroecology grew more rapidly. Gliessman describes that the innovative work of Prof. Efraim Hernandez X., who developed research based on indigenous systems of knowledge in Mexico, led to education programs in agroecology. In 1977 Prof. Efraim Hernandez X. explained that modern agricultural systems had lost their ecological foundation when socio-economic factors became the only driving force in the food system. The acknowledgement that the socio-economic interactions are indeed one of the fundamental components of any agroecosystems came to light in 1982, with the article Agroecologia del Tropico Americano by Montaldo. The author argues that the socio-economic context cannot be separated from the agricultural systems when designing agricultural practices.

In 1995 Edens et al. in Sustainable Agriculture and Integrated Farming Systems solidified this idea proving his point by devoting special sections to economics of the systems, ecological impacts, and ethics and values in agriculture. Actually, 1985 ended up being a fertile and creative year for the new discipline. For instance in the same year, Miguel Altieri integrated how consolidation of the farms, and cropping systems impact pest populations. In addition, Gliessman highlighted that socio-economic, technological, and ecological components give rise to producer choices of food production systems. These pioneering agroecologists have helped to frame the foundation of what we today consider the interdisciplinary field of agroecology.

Publications

Year	Author(s)	Title
1928	Klages	Crop ecology and ecological crop geography in the agronomic curriculum
1939	Hanson	Ecology in agriculture
1956	Azzi	Agricultural ecology
1965	Tischler	Agrarökologie
1973	Janzen	Tropical agroecosystems
1974	Harper	The need for a focus on agro-ecosystems

1976	Loucks	Emergence of research on agroecosystems
1977	Hernanez Xolocotzi	Agroecosistemas de Mexico
1978	Gliessman	Agroecosistemas y tecnologia agricola tradicional
1979	Hart	Agroecosistemas: conceptos básicos
1979	Cox & Atkins	Agricultural ecology: an analysis of world food production systems
1980	Hart	Agroecosistemas
1981	Gliessman, Garcia	The ecological basis for the application
	& Amador	of traditional agricultural technology in the management of tropical agroecosystems
1982	Montaldo	Agroecologia del trópico americano
1983	Altieri	Agroecology
1984	Lowrance, Stinner & House	Agricultural ecosystems: unifying concepts
1985	Conway	Agroecosystems analysis
1987	Altieri	Agroecology: the scientific basis of alternative agriculture
1990	Allen, Dusen, Lundy, & Gliessman	Integrating social, environmental, and economic issues in sustainable agriculture
1990	Gliessman	Agroecology: researching the ecological basis for sustainable agriculture
1990	Carroll, Vandermeer & Rosset	Agroecology
1990	Altieri & Hecht	Agroecology and small farm development
1991	Caporali	Ecologia per l'agricultura
1991	Bawden	Systems thinking in agriculture
1993	Coscia	Agricultura sostenible
1998	Gliessman	Agroecology: ecological processes in sustainable agriculture
2001	Flora	Interactions between agroecosystems and rural communities
2001	Gliessman	Agroecosystem sustainability

2002	Dalgaard, Porter & Hutchings	Agroecology, scaling, and interdisciplinarity
2003	Francis et al.	Agroecology: The Ecology of Food Systems
2004	Clements, Shrestra	New Dimension in Agroecology
2007	Bland and Bell	A Holon Approach to Agroecology
2007	Gliessman	Agroecology: The Ecology of Sustainable Food Systems
2007	Warner	Agroecology in Action
2009	Wezel, Soldat	A quantitative and qualitative historical analysis of the scientific discipline agroecology
2009	Wezel et al.	Agroecology as a science, a movement or a practice. A review

Applications

To emit a point of view about a particular way of farming, an agroecologist would first seek to understand the contexts in which the farm(s) is(are) involved. Each farm may be inserted in a unique combination of factors or contexts. Each farmer may have their own premises about the meanings of an agricultural endeavour, and these meanings might be different than those of agroecologists. Generally, farmers seek a configuration that is viable in multiple contexts, such as family, financial, technical, political, logistical, market, environmental, spiritual. Agroecologists want to understand the behaviour of those who seek livelihoods from plant and animal increase, acknowledging the organisation and planning that is required to run a farm.

Views on Organic and Non-organic Milk Production

Because organic agriculture proclaims to sustain the health of soils, ecosystems, and people, it has much in common with Agroecology; this does not mean that Agroecology is synonymous with organic agriculture, nor that Agroecology views organic farming as the 'right' way of farming. Also, it is important to point out that there are large differences in organic standards among countries and certifying agencies. Three of the main areas that agroecologists would look at in farms, would be: the environmental impacts, animal welfare issues, and the social aspects.

Environmental impacts caused by organic and non-organic milk production can vary significantly. For both cases, there are positive and negative environmental consequences.

Compared to conventional milk production, organic milk production tends to have lower eutrophication potential per ton of milk or per hectare of farmland, because it potentially reduces leaching of nitrates (NO_3^-) and phosphates (PO_4^-) due to lower fertilizer application rates. Because organic milk production reduces pesticides utilisation, it increases land use per ton of milk due to decreased crop yields per hectare. Mainly due to the lower level of concentrates given to cows in organic herds, organic dairy farms generally produce less milk per cow than conventional dairy farms. Because of the increased use of roughage and the, on-average, lower milk production level per cow, some research has connected organic milk production with increases in the emission of methane.

Animal welfare issues vary among dairy farms and are not necessarily related to the way of producing milk (organically or conventionally).

A key component of animal welfare is freedom to perform their innate (natural) behaviour, and this is stated in one of the basic principles of organic agriculture. Also, there are other aspects of animal welfare to be considered- such as freedom from hunger, thirst, discomfort, injury, fear, distress, disease and pain. Because organic standards require loose housing systems, adequate bedding, restrictions on the area of slatted floors, a minimum forage proportion in the ruminant diets, and tend to limit stocking densities both on pasture and in housing for dairy cows, they potentially promote good foot and hoof health. Some studies show lower incidence of placenta retention, milk fever, abomasums displacement and other diseases in organic than in conventional dairy herds. However, the level of infections by parasites in organically managed herds is generally higher than in conventional herds.

Social aspects of dairy enterprises include life quality of farmers, of farm labour, of rural and urban communities, and also includes public health.

Both organic and non-organic farms can have good and bad implications for the life quality of all the different people involved in that food chain. Issues like labour conditions, labour hours and labour rights, for instance, do not depend on the organic/non-organic characteristic of the farm; they can be more related to the socio-economical and cultural situations in which the farm is inserted, instead.

As for the public health or food safety concern, organic foods are intended to be healthy, free of contaminations and free from agents that could cause human diseases. Organic milk is meant to have no chemical residues to consumers, and the restrictions on the use of antibiotics and chemicals in organic food production has the purpose to accomplish this goal. But dairy cows in organic farms, as in conventional farms, indeed do get exposed to virus, parasites and bacteria that can contaminate milk and hence humans, so the risks of transmitting diseases are not eliminated just because the production is organic.

In an organic dairy farm, an agroecologist could evaluate the following:

1. Can the farm minimise environmental impacts and increase its level of sustainability, for instance by efficiently increasing the productivity of the animals to minimise waste of feed and of land use?

2. Are there ways to improve the health status of the herd (in the case of organics, by using biological controls, for instance)?

3. Does this way of farming sustain good quality of life for the farmers, their families, rural labour and communities involved?

Views on No-till Farming

No-tillage is one of the components of conservation agriculture practices and is considered more environmental friendly than complete tillage. Due to this belief, it could be expected that agroecologists would not recommend the use of complete tillage and would rather recommend no-till farming, but this is not always the case. In fact, there is a general consensus that no-till can increase soils capacity of acting as a carbon sink, especially when combined with cover crops.

No-till can contribute to higher soil organic matter and organic carbon content in soils, though reports of no-effects of no-tillage in organic matter and organic carbon soil also exist, depending on environmental and crop conditions. In addition, no-till can indirectly reduce CO_2 emissions by decreasing the use of fossil fuels.

Most crops can benefit from the practice of no-till, but not all crops are suitable for complete no-till agriculture. Crops that do not perform well when competing with other plants that grow in untilled soil in their early stages can be best grown by using other conservation tillage practices, like a combination of strip-till with no-till areas. Also, crops which harvestable portion grows underground can have better results with strip-tillage, mainly in soils which are hard for plant roots to penetrate into deeper layers to access water and nutrients. The benefits provided by no-tillage to predators may lead to larger predator populations, which is a good way to control pests (biological control), but also can facilitate predation of the crop itself. In corn crops, for instance, predation by caterpillars can be higher in no-till than in conventional tillage fields. In places with rigorous winter, untilled soil can take longer to warm and dry in spring, which may delay planting to less ideal dates. Another factor to be considered is that organic residue from the prior year's crops laying on the surface of untilled fields can provide a favourable environment to pathogens, helping to increase the risk of transmitting diseases to the future crop. And because no-till farming provides good environment for pathogens, insects and weeds, it can lead farmers to a more intensive use of chemicals for pest control. Other disadvantages of no-till include underground rot, low soil temperatures and high moisture.

Based on the balance of these factors, and because each farm has different problems, agroecologists will not atest that only no-till or complete tillage is the right way of farming. Yet, these are not the only possible choices regarding soil preparation, since there are intermediate practices such as strip-till, mulch-till and ridge-till, all of them- just as no-till- categorised as conservation tillage. Agroecologists, then, will evaluate the need of different practices for the contexts in which each farm is inserted.

In a no-till system, an agroecologist could ask the following:

1. Can the farm minimise environmental impacts and increase its level of sustainability; for instance by efficiently increasing the productivity of the crops to minimise land use?

2. Does this way of farming sustain good quality of life for the farmers, their families, rural labour and rural communities involved?

By Region

The principles of agroecology are expressed differently depending on local ecological and social contexts.

Latin America

Latin America's experiences with North American Green Revolution agricultural techniques have opened space for agroecologists. Traditional or indigenous knowledge represents a wealth of possibility for agroecologists, including "exchange of wisdoms."

Madagascar

Most of the historical farming in Madagascar has been conducted by indigenous peoples. The French colonial period disturbed a very small percentage of land area, and even included some useful experiments in sustainable forestry. Slash-and-burn techniques, a component of some shifting cultivation systems have been practised by natives in Madagascar for centuries. As of 2006 some of the major agricultural products from slash-and-burn methods are wood, charcoal and grass for Zebu grazing. These practices have taken perhaps the greatest toll on land fertility since the end of French rule, mainly due to overpopulation pressures.

Agriculture

Agriculture (also called farming or husbandry) is the cultivation of animals, plants, fungi and other life forms for food, fibre, and other products used to sustain life. Agriculture was the key implement in the rise of sedentary human civilization, whereby farming of domesticated species created food surpluses that nurtured the development of civilization. The study of agriculture is known as agricultural science. Agriculture is also observed in certain species of ant and termite, but generally speaking refers to human activities.

The history of agriculture dates back thousands of years, and its development has been driven and defined by greatly different climates, cultures, and technologies. However, all farming generally relies on techniques to expand and maintain the lands suitable for raising domesticated species. For plants, this usually requires some form of irrigation, although there are methods of dryland farming; pastoral herding on rangeland is still the most common means of raising livestock. In the developed world, industrial agriculture based on large-scale monoculture has become the dominant system of modern farming, although there is growing support for sustainable agriculture (e.g. permaculture or organic agriculture).

Modern agronomy, plant breeding, pesticides and fertilizers, and technological improvements have sharply increased yields from cultivation, but at the same time have caused widespread ecological damage and negative human health effects. Selective breeding and modern practices in animal husbandry such as intensive pig farming have similarly increased the output of meat, but have raised concerns about animal cruelty and the health effects of the antibiotics, growth hormones, and other chemicals commonly used in industrial meat production.

The major agricultural products can be broadly grouped into foods, fibres, fuels, and raw materials. In the 21st century, plants have been used to grow biofuels, biopharmaceuticals, bioplastics, and pharmaceuticals. Specific foods include cereals, vegetables, fruits, and meat. Fibres include cotton, wool, hemp, silk and flax. Raw materials include lumber and bamboo. Other useful materials are produced by plants, such as resins. Biofuels include methane from biomass, ethanol, and biodiesel. Cut flowers, nursery plants, tropical fish and birds for the pet trade are some of the ornamental products. Regarding food production, the World Bank targets agricultural food production and water management as an increasingly global issue that is fostering an important and growing debate.

In 2007, one third of the world's workers were employed in agriculture. The services sector has overtaken agriculture as the economic sector employing the most people worldwide. Despite the size of its workforce, agricultural production accounts for less than five percent of the gross world product (an aggregate of all gross domestic products).

Etymology

The word *agriculture* is the English adaptation of Latin *agricultûra*, from *ager*, "a field", and *cultûra*, "cultivation" in the strict sense of "tillage of the soil". Thus, a literal reading of the word yields "tillage of a field/of fields".

Overview

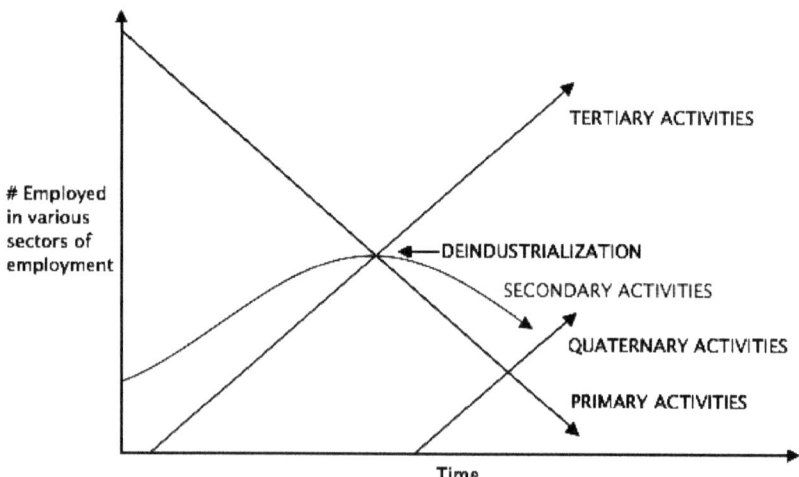

Clark's Sector Model (1950): The percent of the human population working in primary sector activities such as agriculture has decreased over time.

Agriculture has played a key role in the development of human civilization. Until the Industrial Revolution, the vast majority of the human population laboured in agriculture. The type of agriculture they developed was typically subsistence agriculture in which farmers raised most of their crops for consumption on farm, and there was only a small portion left over for the payment of taxes, dues, or trade. In subsistence agriculture cropping decisions are made with an eye to what the family needs for food, and to make clothing, and not the world marketplace. Development of agricultural techniques has steadily increased agricultural productivity, and the widespread diffusion of these techniques during a time period is often called an agricultural revolution. A remarkable shift in agricultural practices has occurred over the past century in response to new technologies, and the development of world markets. This also led to technological improvements in agricultural techniques, such as the Haber-Bosch method for synthesizing ammonium nitrate which made the traditional practice of recycling nutrients with crop rotation and animal manure less necessary.

Synthetic nitrogen, along with mined rock phosphate, pesticides and mechanisation, have greatly increased crop yields in the early 20th century. Increased supply of grains has led to cheaper livestock as well. Further, global yield increases were experienced later in the 20th century when high-yield varieties of common staple grains such as rice, wheat, and corn (maize) were introduced as a part of the Green Revolution. The Green Revolution exported the technologies (including pesticides and synthetic nitrogen) of the developed world to the developing world. Thomas Malthus famously predicted that the Earth would not be able to support its growing population, but technologies such as the Green Revolution have allowed the world to produce a surplus of food.

Many governments have subsidised agriculture to ensure an adequate food supply. These agricultural subsidies are often linked to the production of certain commodities such as wheat, corn (maize), rice, soybeans, and milk. These subsidies, especially when instituted by developed countries have been noted as protectionist, inefficient, and environmentally damaging.

In the past century agriculture has been characterised by enhanced productivity, the use of synthetic fertilizers and pesticides, selective breeding, mechanisation, water contamination, and farm subsidies. Proponents of organic farming such as Sir Albert Howard argued in the early 20th century that the overuse of pesticides and synthetic fertilizers damages the long-term fertility of the soil. While this feeling lay dormant for decades, as environmental awareness has increased in the 21st century there has been a movement towards sustainable agriculture by some farmers, consumers, and policymakers.

In recent years there has been a backlash against perceived external environmental effects of mainstream agriculture, particularly regarding water pollution, resulting in the organic movement. One of the major forces behind this movement has been the European Union, which first certified organic food in 1991 and began reform of its Common Agricultural Policy(CAP) in 2005 to phase out commodity-linked farm subsidies, also known as decoupling. The growth of organic farming has renewed research in alternative technologies such as integrated pest management and selective breeding. Recent mainstream technological developments include genetically modified food.

In late 2007, several factors pushed up the price of grains consumed by humans as well as used to feed poultry and dairy cows and other cattle, causing higher prices of wheat (up 58%), soybean (up 32%), and maize (up 11%) over the year. Food riots took place in several countries across the world. Contributing factors included drought in Australia and elsewhere, increasing demand for grain-fed animal products from the growing middle classes of countries such as China and India, diversion of foodgrain to biofuel production and trade restrictions imposed by several countries. An epidemic of stem rust on wheat caused by race Ug99 is currently spreading across Africa and into Asia and is causing major concern. Approximately 40% of the world's agricultural land is seriously degraded. In Africa, if current trends of soil degradation continue, the continent might be able to feed just 25% of its population by 2025, according to UNU's Ghana-based Institute for Natural Resources in Africa.

History

Agricultural practices such as irrigation, crop rotation, fertilizers, and pesticides were developed long ago, but have made great strides in the past century. The history of agriculture has played a major role in human history, as agricultural progress has been a crucial factor in worldwide socio-economic change. Division of labour in agricultural societies made commonplace specialisations rarely seen inhunter-gatherer cultures. So, too, are arts such as epic literature and monumental architecture, as well as codified legal systems. When farmers became capable of producing food beyond the needs of their own families, others in their society were freed to devote themselves to projects other than food acquisition. Historians and anthropologists have long argued that the development of agriculture made civilization possible. The total world population probably never exceeded 15 million inhabitants before the invention of agriculture.

Ancient Origins

The Fertile Crescent of Western Asia, Egypt, and India were sites of the earliest planned sowing and harvesting of plants that had previously been gathered in the wild. Independent development of agriculture occurred in northern and southern China, Africa's Sahel, New Guinea and several regions of the Americas. The eight so-called Neolithic founder crops of agriculture appear: first emmer wheat and einkorn wheat, then hulled barley, peas, lentils, bitter vetch, chick peas and flax.

By 7000 BC, small-scale agriculture reached Egypt. From at least 7000 BC the Indian subcontinent saw farming of wheat and barley, as attested by archaeological excavation at Mehrgarh inBalochistan in what is present day Pakistan. By 6000 BC, mid-scale farming was entrenched on the banks of the Nile. This, as irrigation had not yet matured sufficiently. About this time, agriculture was developed independently in the Far East, with rice, rather than wheat, as the primary crop. Chinese and Indonesian farmers went on to domesticate taro and beans including mung, soy andazuki. To complement these new sources of carbohydrates, highly organised net fishing of rivers, lakes and ocean shores in these areas brought in great volumes of essential protein. Collectively, these new methods of farming and fishing inaugurated a human population boom that dwarfed all previous expansions and continues today.

By 5000 BC, the Sumerians had developed core agricultural techniques including large-scale intensive cultivation of land, monocropping, organised irrigation, and the use of a specialised labour force, particularly along the waterway now known as the Shatt al-Arab, from its Persian Gulf delta to the confluence of the Tigris and Euphrates. Domestication of wild aurochs and mouflon into cattle and sheep, respectively, ushered in the large-scale use of animals for food/fibre and as beasts of burden. The shepherd joined the farmer as an essential provider for sedentary and seminomadic societies. Maize, manioc, and arrowroot were first domesticated in the Americas as far back as 5200 BC.

The potato, tomato, pepper, squash, several varieties of bean, tobacco, and several other plants were also developed in the Americas, as was extensive terracing of steep hillsides in much of Andean South America. The Greeks and Romansbuilt on techniques pioneered by the Sumerians, but made few fundamentally new advances. Southern Greeks struggled with very poor soils, yet managed to become a dominant society for years. The Romans were noted for an emphasis on the cultivation of crops for trade. In the same region, a parallel agricultural revolution occurred, resulting in some of the most important crops grown today.

In Mesoamerica wild teosinte was transformed through human selection into the ancestor of modern maize, more than 6000 years ago. It gradually spread across North America and was the major crop of Native Americans at the time of European exploration. Other Mesoamerican crops include hundreds of varieties of squash and beans.

Cocoa was also a major crop in domesticated Mexico and Central America. The turkey, one of the most important meat birds, was probably domesticated in Mexico or the U.S. Southwest. In the Andes region of South America the major domesticated crop was potatoes, domesticated perhaps 5000 years ago. Large varieties of beans were domesticated, in South America, as well as animals, including llamas, alpacas, and guinea pigs. Coca, still a major crop, was also domesticated in the Andes.

A minor centre of domestication, the indigenous people of the Eastern U.S. appear to have domesticated numerous crops. Sunflowers, tobacco, varieties of squash and Chenopodium, as well as crops no longer grown, including marshelder and little barley were domesticated. Other wild foods may have undergone some selective cultivation, including wild rice and maple sugar. The most common varieties of strawberry were domesticated from Eastern North America.

By 3500 BC, the simplest form of the plough was developed, called the ard. Before this period, simple digging sticks or hoes were used. These tools would have also been easier to transport, which was a benefit as people only stayed until the soil's nutrients were depleted. However, through excavations in Mexico it has been found that the continuous cultivating of smaller pieces of land would also have been a sustaining practice. Additional research in central Europe later revealed that agriculture was indeed practiced at this method. For this method, ards were thus much more efficient than digging sticks.

Middle Ages

During the Middle Ages, farmers in North Africa, the Near East, and Europe began making use of agricultural technologies including irrigation systems based on hydraulic and hydrostatic principles, machines such as norias, water-raising machines, dams, and reservoirs. This combined with the invention of a three-field system of crop rotation and the moldboard plow greatly improved agricultural efficiency.

In the European medieval period, agriculture was considered part of the set of *seven mechanical arts*. "Between 1413 and 1635, the top 5 percent of villagers tripled the amount of arable land they held, (Duplessis). The wealth getting wealthier.

Modern Era

The Harvesters. Pieter Bruegel. 1565.

After 1492, a global exchange of previously local crops and livestock breeds occurred. Key crops involved in this exchange included the tomato, maize, potato, manioc, cocoa bean and tobacco going from the New World to the Old, and several varieties of wheat, spices, coffee, and sugar cane going from the Old World to the New.

The most important animal exportation from the Old World to the New were those of the horse and dog (dogs were already present in the pre-Columbian Americas but not in the numbers and breeds suited to farm work). Although not usually food animals, the horse (including donkeys and ponies) and dog quickly filled essential production roles on western-hemisphere farms.

The potato became an important staple crop in northern Europe. Since being introduced by Portuguese in the 16th century, maize and manioc have replaced traditional African crops as the continent's most important staple food crops.

By the early 19th century, agricultural techniques, implements, seed stocks and cultivar had so improved that yield per land unit was many times that seen in the Middle Ages. Although there is a vast and interesting history of crop cultivation before the dawn of the 20th century, there is little question that the work of Charles Darwin and Gregor Mendel created the scientific foundation for plant breeding that led to its explosive impact over the past 150 years.

With the rapid rise of mechanisation in the late 19th century and the 20th century, particularly in the form of the tractor, farming tasks could be done with a speed and on a scale previously impossible. These advances have led to efficiencies enabling certain modern farms in the United States, Argentina, Israel, the United Kingdom Germany, and a few other nations to output volumes of high-quality produce per land unit at what may be the practical limit.

The Haber-Bosch method for synthesizing ammonium nitrate represented a major breakthrough and allowed crop yields to overcome previous constraints. In the past century agriculture has been characterised by enhanced productivity, the substitution of synthetic fertilizers and pesticides for labour, water pollution, and farm subsidies. In recent years there has been a backlash against the external environmental effects of conventional agriculture, resulting in the organic movement.

The cereals rice, corn, and wheat provide 60% of human food supply. Between 1700 and 1980, "the total area of cultivated land worldwide increased 466%" and yields increased dramatically, particularly because of selectively bred high-yielding varieties, fertilizers, pesticides, irrigation, and machinery. For example, irrigation increased corn yields in eastern Colorado by 400 to 500% from 1940 to 1997.

However, concerns have been raised over the sustainability of intensive agriculture. Intensive agriculture has become associated with decreased soil quality in India and Asia, and there has been increased concern over the effects of fertilizers and pesticides on the environment, particularly as population increases and food demand expands. The monocultures typically used in intensive agriculture increase the number of pests, which are controlled through pesticides. Integrated pest management (IPM), which "has been promoted for decades and has had some notable successes" has not significantly affected the use of pesticides because policies encourage the use of pesticides and IPM is knowledge-intensive.

Although the "Green Revolution" significantly increased rice yields in Asia, yield increases have not occurred in the past 15–20 years. The genetic "yield potential" has increased for wheat, but the yield potential for rice has not increased since 1966, and the yield potential for maize has "barely increased in 35 years". It takes a decade or two for herbicide-resistant weeds to emerge, and insects become resistant to insecticides within about a decade. Crop rotation helps to prevent resistances. Agricultural exploration expeditions, since the late 19th century, have been mounted to find new species and new agricultural practices in different areas of the world. Two early examples of expeditions include Frank N. Meyer's fruit- and nut-collecting trip to China and Japan from 1916-1918 and the Dorsett-Morse Oriental Agricultural Exploration Expedition to China, Japan, and Korea from 1929-1931 to collect soybean germplasm to support the rise in soybean agriculture in the United States.

In 2009, the agricultural output of China was the largest in the world, followed by the European Union, India and the United States, according to the International Monetary Fund. Economists measure the total factor productivity of agriculture and by this measure agriculture in the United States is roughly 2.6 times more productive than it was in 1948. Six countries- the US, Canada, France, Australia, Argentina and Thailand- supply 90% of grain exports. Water deficits, which are already spurring heavy grain imports in numerous middle-sized countries, including Algeria, Iran, Egypt, and Mexico, may soon do the same in larger countries, such as China or India.

Crop Production Systems

Banaue Rice Terraces, Ifugao Province,Philippines

Cropping systems vary among farms depending on the available resources and constraints; geography and climate of the farm; government policy; economic, social and political pressures; and the philosophy and culture of the farmer. Shifting cultivation (or slash and burn) is a system in which forests are burnt, releasing nutrients to support cultivation of annual and then perennial crops for a period of several years.

Then the plot is left fallow to regrow forest, and the farmer moves to a new plot, returning after many more years (10-20). This fallow period is shortened if population density grows, requiring the input of nutrients (fertilizer or manure) and some manual pest control. Annual cultivation is the next phase of intensity in which there is no fallow period. This requires even greater nutrient and pest control inputs.

Further industrialisation lead to the use of monocultures, when one cultivar is planted on a large acreage. Because of the low biodiversity, nutrient use is uniform and pests tend to build up, necessitating the greater use of pesticides and fertilizers. Multiple cropping, in which several crops are grown sequentially in one year, and intercropping, when several crops are grown at the same time are other kinds of annual cropping systems known as polycultures.

In tropical environments, all of these cropping systems are practiced. In subtropical and arid environments, the timing and extent of agriculture may be limited by rainfall, either not allowing multiple annual crops in a year, or requiring irrigation. In all of these environments perennial crops are grown (coffee, chocolate) and systems are practiced such as agroforestry. In temperate environments, where ecosystems were predominantly grassland or prairie, highly productive annual cropping is the dominant farming system.

The last century has seen the intensification, concentration and specialisation of agriculture, relying upon new technologies of agricultural chemicals (fertilizers and pesticides), mechanisation, and plant breeding (hybrids and GMO's).

In the past few decades, a move towards sustainability in agriculture has also developed, integrating ideas of socio-economic justice and conservation of resources and the environment within a farming system. This has led to the development of many responses to the conventional agriculture approach, including organic agriculture, urban agriculture, community supported agriculture, ecological or biological agriculture, integrated farming and holistic management, as well as an increased trend towards agricultural diversification.

Crop Statistics

Important categories of crops include grains and pseudograins, pulses (legumes), forage, and fruits and vegetables. Specific crops are cultivated in distinct growing regions throughout the world. In millions of metric tons, based on FAO estimate.

Top agricultural products, by crop types(million tonnes) 2004 data

Cereals	2,263
Vegetables and melons	866
Roots and Tubers	715
Milk	619
Fruit	503
Meat	259
Oilcrops	133
Fish (2001 estimate)	130
Eggs	63
Pulses	60
Vegetable Fibre	30

Source: Food and Agriculture Organisation (FAO)

Top agricultural products, by individual crops(million tonnes) 2004 data

Sugar Cane	1,324
Maize	721
Wheat	627
Rice	605
Potatoes	328
Sugar Beet	249
Soybean	204
Oil Palm Fruit	162
Barley	154
Tomato	120

Source: Food and Agriculture Organisation (FAO)

Livestock Production Systems

Animals, including horses, mules, oxen, camels, llamas, alpacas, and dogs, are often used to help cultivate fields, harvest crops, wrangle other animals, and transport farm products to buyers. Animal husbandry not only refers to the breeding and raising of animals for meat or to harvest animal products (like milk, eggs, or wool) on a continual basis, but also to the breeding and care of species for work and companionship. Livestock production systems can be defined based on feed source, as grassland- based, mixed, and landless.

Grassland based livestock production relies upon plant material such as shrubland, rangeland, and pastures for feeding ruminant animals. Outside nutrient inputs may be used, however manure is returned directly to the grassland as a major nutrient source. This system is particularly important in areas where crop production is not feasible because of climate or soil, representing 30-40 million pastoralists. Mixed production systems use grassland, fodder crops and grain feed crops as feed for ruminant and monogastic (one stomach; mainly chickens and pigs) livestock. Manure is typically recycled in mixed systems as a fertilizer for crops. Approximately 68% of all agricultural land is permanent pastures used in the production of livestock.

Ploughing rice paddies with water buffalo, in Indonesia.

Landless systems rely upon feed from outside the farm, representing the de-linking of crop and livestock production found more prevalently in OECD member countries. In the U.S., 70% of the grain grown is fed to animals on feedlots. Synthetic fertilizers are more heavily relied upon for crop production and manure utilisation becomes a challenge as well as a source for pollution.

Production Practices

Tillage is the practice of plowing soil to prepare for planting or for nutrient incorporation or for pest control. Tillage varies in intensity from conventional to no-till. It may improve productivity by warming the soil, incorporating fertilizer and controlling weeds, but also renders soil more prone to erosion, triggers the decomposition of organic matter releasing CO_2, and reduces the abundance and diversity of soil organisms.

Pest control includes the management of weeds, insects/mites, and diseases. Chemical (pesticides), biological (biocontrol), mechanical (tillage), and cultural practices are used. Cultural practices include crop rotation, culling, cover crops, intercropping, composting, avoidance, and resistance. Integrated pest management attempts to use all of these methods to keep pest populations below the number which would cause economic loss, and recommends pesticides as a last resort.

Road leading across the farm allows machinery access to the farm for production practices.

Nutrient management includes both the source of nutrient inputs for crop and livestock production, and the method of utilisation of manure produced by livestock. Nutrient inputs can be chemical inorganic fertilizers, manure, green manure, compost and mined minerals. Crop nutrient use may also be managed using cultural techniques such as crop rotation or a fallow period. Manure is used either by holding livestock where the feed crop is growing, such as in managed intensive rotational grazing, or by spreading either dry or liquid formulations of manure on cropland or pastures.

Water management is where rainfall is insufficient or variable, which occurs to some degree in most regions of the world. Some farmers use irrigation to supplement rainfall. In other areas such as the Great Plains in the U.S. and Canada, farmers use a fallow year to conserve soil moisture to use for growing a crop in the following year. Agriculture represents 70% of freshwater use worldwide.

Processing, Distribution, and Marketing

In the United States, food costs attributed to processing, distribution, and marketing have risen while the costs attributed to farming have declined. This is related to the greater efficiency of farming, combined with the increased level of value addition (e.g. more highly processed products) provided by the supply chain. From 1960 to 1980 the farm share was around 40%, but by 1990 it had declined to 30% and by 1998, 22.2%. Market concentration has increased in the sector as well, with the top 20 food manufacturers accounting for half the food-processing value in 1995, over double that produced in 1954. As of 2000 the top six US supermarket groups had 50% of sales compared to 32% in 1992. Although the total effect of the increased market concentration is likely increased efficiency, the changes redistribute economic surplus from producers (farmers) and consumers, and may have negative implications for rural communities.

Crop Alteration and Biotechnology

Tractor and Chaser bin.

Crop alteration has been practiced by humankind for thousands of years, since the beginning of civilization. Altering crops through breeding practices changes the genetic make-up of a plant to develop crops with more beneficial characteristics for humans, for example, larger fruits or seeds, drought-tolerance, or resistance to pests. Significant advances in plant breeding ensued after the work of geneticist Gregor Mendel. His work on dominant and recessive alleles gave plant breeders a better understanding of genetics and brought great insights to the techniques utilised by plant breeders. Crop breeding includes techniques such as plant selection with desirable traits, self-pollination and cross-pollination, and molecular techniques that genetically modify the organism. Domestication of plants has, over the centuries increased yield, improved disease resistance and drought tolerance, eased harvest and improved the taste and nutritional value of crop plants. Careful selection and breeding have had enormous effects on the characteristics of crop plants. Plant selection and breeding in the 1920s and 1930s improved pasture (grasses and clover) in New Zealand. Extensive X-ray and ultraviolet induced mutagenesis efforts (i.e. primitive genetic engineering) during the 1950s produced the modern commercial varieties of grains such as wheat, corn (maize) and barley.

The Green Revolution popularised the use of conventional hybridisation to increase yield many folds by creating "high-yielding varieties". For example, average yields of corn (maize) in the USA have increased from around 2.5 tons per hectare (t/ha) (40 bushels per acre) in 1900 to about 9.4 t/ha (150 bushels per acre) in 2001. Similarly, worldwide average wheat yields have increased from less than 1 t/ha in 1900 to more than 2.5 t/ha in 1990. South American average wheat yields are around 2 t/ha, African under 1 t/ha, Egypt and Arabia up to 3.5 to 4 t/ha with irrigation. In contrast, the average wheat yield in countries such as France is over 8 t/ha. Variations in yields are due mainly to variation in climate, genetics, and the level of intensive farming techniques (use of fertilizers, chemical pest control, growth control to avoid lodging).

Genetic Engineering

Genetically Modified Organisms (GMO) are organisms whose genetic material has been altered by genetic engineering techniques generally known as recombinant DNA technology. Genetic engineering has expanded the genes available to breeders to utilise in creating desired germlines for new crops. After mechanical tomato-harvesters were developed in the early 1960s, agricultural scientists genetically modified tomatoes to be more resistant to mechanical handling. More recently, genetic engineering is being employed in various parts of the world, to create crops with other beneficial traits. New research on woodland strawberry genome was found to be short and easy to manipulate. Researchers now have tools to improve strawberry flavours and aromas of cultivated strawberries as stated in a publication by Nature Genetics.

Herbicide-tolerant GMO Crops

Roundup Ready seed has a herbicide resistant gene implanted into its genome that allows the plants to tolerate exposure to glyphosate. *Roundup* is a trade name for a glyphosate-based product, which is a systemic, nonselective herbicide used to kill weeds. *Roundup Ready* seeds allow the farmer to grow a crop that can be sprayed with glyphosate to control weeds without harming the resistant crop. Herbicide-tolerant crops are used by farmers worldwide. Today, 92% of soybean acreage in the US is planted with genetically modified herbicide-tolerant plants.

With the increasing use of herbicide-tolerant crops, comes an increase in the use of glyphosate-based herbicide sprays. In some areas glyphosate resistant weeds have developed, causing farmers to switch to other herbicides. Some studies also link widespread glyphosate usage to iron deficiencies in some crops, which is both a crop production and a nutritional quality concern, with potential economic and health implications.

Insect-resistant GMO Crops

Other GMO crops used by growers include insect-resistant crops, which have a gene from the soil bacterium *Bacillus thuringiensis* (Bt), which produces a toxin specific to insects. These crops protect plants from damage by insects; one such crop is Starlink. Another is cotton, which accounts for 63% of US cotton acreage.

Some believe that similar or better pest-resistance traits can be acquired through traditional breeding practices, and resistance to various pests can be gained through hybridisation or cross-pollination with wild species. In some cases, wild species are the primary source of resistance traits; some tomato cultivars that have gained resistance to at least 19 diseases did so through crossing with wild populations of tomatoes.

Costs and Benefits of GMOs

Genetic engineers may someday develop transgenic plants which would allow for irrigation, drainage, conservation, sanitary engineering, and maintaining or increasing yields while requiring fewer fossil fuel derived inputs than conventional crops. Such developments would be particularly important in areas which are normally arid and rely upon constant irrigation, and on large scale farms. However, genetic engineering of plants has proven to be controversial. Many issues surrounding food security and environmental impacts have risen regarding GMO practices. For example, GMOs are questioned by some ecologists and economists concerned with GMO practices such as terminator seeds, which is a genetic modification that creates sterile seeds. Terminator seeds are currently under strong international opposition and face continual efforts of global bans.

Another controversial issue is the patent protection given to companies that develop new types of seed using genetic engineering. Since companies have intellectual ownership of their seeds, they have the power to dictate terms and conditions of their patented product. Currently, ten seed companies control over two-thirds of the global seed sales.

Vandana Shiva argues that these companies are guilty of biopiracy by patenting life and exploiting organisms for profit Farmers using patented seed are restricted from saving seed for subsequent plantings, which forces farmers to buy new seed every year. Since seed saving is a traditional practice for many farmers in both developing and developed countries, GMO seeds legally bind farmers to change their seed saving practices to buying new seed every year.

Locally adapted seeds are an essential heritage that has the potential to be lost with current hybridised crops and GMOs. Locally adapted seeds, also called land races or crop eco-types, are important because they have adapted over time to the specific micro-climates, soils, other environmental conditions, field designs, and ethnic preference indigenous to the exact area of cultivation.

Introducing GMOs and hybridised commercial seed to an area brings the risk of cross-pollination with local land races Therefore, GMOs pose a threat to the sustainability of land races and the ethnic heritage of cultures. Once seed contains transgenic material, it becomes subject to the conditions of the seed company that owns the patent of the transgenic material.

Modern Agriculture

Modern agriculture is a term used to describe the wide majority of production practices employed by America's farmers. The term depicts the push for innovation, stewardship and advancements continually made by growers to sustainability produce higher-quality products with a reduced environmental impact. Intensive scientific research and robust investment in modern agriculture during the past 50 years has helped farmers double food production.

Safety

The agriculture industry works with government agencies and other organisations to ensure that farmers have access to the technologies required to support modern agriculture practices. Farmers are supported by education and certification programs that ensure they apply agricultural practices with care and only when required.

Sustainability

Technological advancements help provide farmers with tools and resources to make farming more sustainable.

New technologies have given rise to innovations like conservation tillage, a farming process which helps prevent land loss to erosion, water pollution and enhances carbon sequestration.

Affordability

The goal of modern agriculture practices is to help farmers provide an affordable supply of food to meet the demands of a growing population. With modern agriculture, more crops can be grown on less land allowing farmers to provide an increased supply of food at an affordable price.

Food Safety, Labelling and Regulation

Food security issues also coincide with food safety and food labelling concerns. Currently a global treaty, the BioSafety Protocol, regulates the trade of GMOs. The EU currently requires all GMO foods to be labelled, whereas the US does not require transparent labelling of GMO foods. Since there are still questions regarding the safety and risks associated with GMO foods, some believe the public should have the freedom to choose and know what they are eating and require all GMO products to be labelled.

The Food and Agriculture Organisation of the United Nations (FAO) leads international efforts to defeat hunger and provides a neutral forum where nations meet as equals to negotiate agreements and debate food policy and the regulation of agriculture. According to Dr. Samuel Jutzi, director of FAO's animal production and health division, lobbying by "powerful" big food corporations has stopped reforms that would improve human health and the environment. The "real, true issues are not being addressed by the political process because of the influence of lobbyists, of the true powerful entities," he said, speaking at the Compassion in World Farming annual forum. For example, recent proposals for a voluntary code of conduct for the livestock industry that would have provided incentives for improving standards for health, and environmental regulations, such as the number of animals an area of land can support without long-term damage, were successfully defeated due to large food company pressure.

Environmental Impact

Agriculture imposes external costs upon society through pesticides, nutrient runoff, excessive water usage, and assorted other problems.

A 2000 assessment of agriculture in the UK determined total external costs for 1996 of £2,343 million, or £208 per hectare. A 2005 analysis of these costs in the USA concluded that cropland imposes approximately $5 to 16 billion ($30 to $96 per hectare), while livestock production imposes $714 million. Both studies concluded that more should be done to internalize external costs, and neither included subsidies in their analysis, but noted that subsidies also influence the cost of agriculture to society. Both focused on purely fiscal impacts. The 2000 review included reported pesticide poisonings but did not include speculative chronic effects of pesticides, and the 2004 review relied on a 1992 estimate of the total impact of pesticides.

In 2010, the International Resource Panel of the United Nations Environment Programme published a report assessing the environmental impacts of consumption and production. The study found that agriculture and food consumption are two of the most important drivers of environmental pressures, particularly habitat change, climate change, water use and toxic emissions.

Agriculture accounts for 70 per cent of withdrawals of freshwater resources. However, increasing pressure being placed on water resources by industry, cities and the involving biofuels industry means that water scarcity is increasing and agriculture is facing the challenge of producing more food for the world's growing population with fewer water resources. Scientists are also realising that water resources need to be allocated to maintain natural environmental services, such as protecting towns from flooding, cleaning ecosystems and supporting fish stocks. In the book *Out of Water: From abundance to scarcity and how to solve the world's water problems*, authors Colin Chartres and Samyukta Varma of the International Water Management Institute lay down a six-point plan of action for addressing the global challenge of producing sufficient food for the world with dwindling water resources. One of the actions they say is required is to ensure all water systems, such as lakes and rivers, have water allocated to environmental flow.

A key player who is credited to saving billions of lives because of his revolutionary work in developing new agricultural techniques is Norman Borlaug. His transformative work brought high-yield crop varieties to developing countries and earned him an unofficial title as the father of the Green Revolution.

Livestock Issues

A senior UN official and co-author of a UN report detailing this problem, Henning Steinfeld, said "Livestock are one of the most significant contributors to today's most serious environmental problems". Livestock production occupies 70% of all land used for agriculture, or 30% of the land surface of the planet. It is one of the largest sources of greenhouse gases, responsible for 18% of the world's greenhouse gas emissions as measured in CO_2 equivalents. By comparison, all transportation emits 13.5% of the CO_2. It produces 65% of human-related nitrous oxide (which has 296 times the global warming potential of CO_2) and 37% of all human-induced methane (which is 23 times as warming as CO_2. It also generates 64% of the ammonia emission. Livestock expansion is cited as a key factor driving deforestation, in the Amazon basin 70% of previously forested area is now occupied by pastures and the remainder used for feedcrops. Through deforestation and land degradation, livestock is also driving reductions in biodiversity.

Land Transformation and Degradation

Land transformation, the use of land to yield goods and services, is the most substantial way humans alter the Earth's ecosystems, and is considered the driving force in the loss of biodiversity. Estimates of the amount of land transformed by humans vary from 39–50%. Land degradation, the long-term decline in ecosystem function and productivity, is estimated to be occurring on 24% of land worldwide, with cropland overrepresented. The UN-FAO report cites land management as the driving factor behind degradation and reports that 1.5 billion people rely upon the degrading land. Degradation can be deforestation, desertification, soil erosion, mineral depletion, or chemical degradation (acidification and salinization).

Eutrophication

Eutrophication, excessive nutrients in aquatic ecosystems resulting in algal blooms and anoxia, leads to fish kills, loss of biodiversity, and renders water unfit for drinking and other industrial uses. Excessive fertilization and manure application to cropland, as well as high livestock stocking densities cause nutrient (mainly nitrogen and phosphorus) runoff and leaching from agricultural land. These nutrients are major nonpoint pollutants contributing to eutrophication of aquatic ecosystems.

Pesticides

Pesticide use has increased since 1950 to 2.5 million tons annually worldwide, yet crop loss from pests has remained relatively constant. The World Health Organisation estimated in 1992 that 3 million pesticide poisonings occur annually, causing 220,000 deaths. Pesticides select for pesticide resistance in the pest population, leading to a condition termed the 'pesticide treadmill' in which pest resistance warrants the development of a new pesticide.

An alternative argument is that the way to 'save the environment' and prevent famine is by using pesticides and intensive high yield farming, a view exemplified by a quote heading the Centre for Global Food Issues website: 'Growing more per acre leaves more land for nature'. However, critics argue that a trade-off between the environment and a need for food is not inevitable, and that pesticides simply replace good agronomic practices such as crop rotation.

Climate Change

Climate change has the potential to affect agriculture through changes in temperature, rainfall (timing and quantity), CO_2, solar radiation and the interaction of these elements. Agriculture can both mitigate or worsen global warming. Some of the increase in CO_2 in the atmosphere comes from the decomposition of organic matter in the soil, and much of the methane emitted into the atmosphere is caused by the decomposition of organic matter in wet soils such as rice paddies. Further, wet or anaerobic soils also lose nitrogen through denitrification, releasing the greenhouse gases nitric oxide and nitrous oxide. Changes in management can reduce the release of these greenhouse gases, and soil can further be used to sequester some of the CO_2 in the atmosphere.

International Economics and Market Reports

Differences in economic development, population density and culture mean that the farmers of the world operate under very different conditions.

A US cotton farmer may receive US\$230 in government subsidies per acre planted (in 2003), while farmers in Mali and other third-world countries do without. When prices decline, the heavily subsidised US farmer is not forced to reduce his output, making it difficult for cotton prices to rebound, but his Mali counterpart may go broke in the meantime.

A livestock farmer in South Korea can calculate with a (highly subsidised) sales price of US\$1300 for a calf produced. A South American Mercosur country rancher calculates with a calf's sales price of US\$120–200 (both 2008 figures). With the former, scarcity and high cost of land is compensated with public subsidies, the latter compensates absence of subsidies with economics of scale and low cost of land.

In the Peoples Republic of China, a rural household's productive asset may be one hectare of farmland. In Brazil, Paraguay and other countries where local legislature allows such purchases, international investors buy thousands of hectares of farmland or raw land at prices of a few hundred US$ per hectare.

To promote exports of agricultural products, many government agencies publish on the web economic studies and reports categorised by product and country. Among these agencies include four of the largest exporters of agricultural products, such as the FAS of the United States Department of Agriculture, Agriculture and Agri-Food Canada (AAFC), Austrade, and NZTE. The Federation of International Trade Associations publishes studies and reports by FAS and AAFC, as well as other non-governmental organisations on its website GlobalTrade.net.

List of Countries by Agricultural Output

Below is a list of countries by agricultural output in 2010.

Table : Agricultural output in 2010 (Nominal)

Rank	Country	Output in billions of US$
—	World	3,585.829
1	China	599.582
—	European Union	293.080
2	India	284.524
3	United States	161.236
4	Brazil	142.141
5	Indonesia	108.130
6	Japan	76.424
7	Turkey	71.218
8	Nigeria	65.041
9	Russia	58.603
10	France	51.651

Table : Agricultural output in 2010 (PPP)

Rank	Country	Output in billions of US$
—	World	4,233.098
1	China	1,028.742
2	India	751.173
—	European Union	273.068
3	United States	161.236

4	Indonesia	157.572
5	Brazil	147.700
6	Nigeria	113.385
7	Pakistan	101.348
8	Turkey	92.209
9	Iran	90.052
10	Russia	88.918

Energy and Agriculture

Since the 1940s, agricultural productivity has increased dramatically, due largely to the increased use of energy-intensive mechanisation, fertilizers and pesticides. The vast majority of this energy input comes from fossil fuel sources. Between 1950 and 1984, the Green Revolution transformed agriculture around the globe, with world grain production increasing by 250% as world population doubled. Modern agriculture's heavy reliance on petrochemicals and mechanisation has raised concerns that oil shortages could increase costs and reduce agricultural output, causing food shortages.

Table : Agriculture and food system share (%) of total energyconsumption by three industrialised nations

Country	Year	Agriculture (direct & indirect)	Foodsystem
United Kingdom	2005	1.9	11
United States of America	1996	2.1	10
United States of America	2002	2.0	14
Sweden	2000	2.5	13

Modern or industrialised agriculture is dependent on fossil fuels in two fundamental ways: 1) direct consumption on the farm and 2) indirect consumption to manufacture inputs used on the farm. Direct consumption includes the use of lubricants and fuels to operate farm vehicles and machinery; and use of gas, liquid propane, and electricity to power dryers, pumps, lights, heaters, and coolers. American farms directly consumed about 1.2 exajoules (1.1 quadrillion BTU) in 2002, or just over 1 percent of the nation's total energy.

Indirect consumption is mainly oil and natural gas used to manufacture fertilizers and pesticides, which accounted for 0.6 exajoules (0.6 quadrillion BTU) in 2002. The energy used to manufacture farm machinery is also a form of indirect agricultural energy consumption, but it is not included in USDA estimates of U.S. agricultural energy use. Together, direct and indirect consumption by

U.S. farms accounts for about 2 percent of the nation's energy use. Direct and indirect energy consumption by U.S. farms peaked in 1979, and has gradually declined over the past 30 years.

Food systems encompass not just agricultural production, but also off-farm processing, packaging, transporting, marketing, consumption, and disposal of food and food-related items. Agriculture accounts for less than one-fifth of food system energy use in the United States.

In 2007, higher incentives for farmers to grow non-food biofuel crops combined with other factors (such as over-development of former farm lands, rising transportation costs, climate change, growing consumer demand in China and India, and population growth) to cause food shortages in Asia, the Middle East, Africa, and Mexico, as well as rising food prices around the globe. As of December 2007, 37 countries faced food crises, and 20 had imposed some sort of food-price controls. Some of these shortages resulted in food riots and even deadly stampedes.

The biggest fossil fuel input to agriculture is the use of natural gas as a hydrogen source for the Haber-Bosch fertilizer-creation process. Natural gas is used because it is the cheapest currently available source of hydrogen. When oil production becomes so scarce that natural gas is used as a partial stopgap replacement, and hydrogen use in transportation increases, natural gas will become much more expensive. If the Haber Process is unable to be commercialised using renewable energy (such as by electrolysis) or if other sources of hydrogen are not available to replace the Haber Process, in amounts sufficient to supply transportation and agricultural needs, this major source of fertilizer would either become extremely expensive or unavailable. This would either cause food shortages or dramatic rises in food prices.

Mitigation of Effects of Petroleum Shortages

In the event of a petroleum shortage, organic agriculture can be more attractive than conventional practices that use petroleum-based pesticides, herbicides, or fertilizers. Some farmers using modern organic-farming methods have reported yields as high as those available from conventional farming. Organic farming may however be more labour-intensive and would require a shift of the workforce from urban to rural areas. The reconditioning of soil to restore nutrients lost during the use of monoculture agriculture techniques also takes time.

It has been suggested that rural communities might obtain fuel from the biochar and synfuel process, which uses agricultural *waste* to provide charcoal fertilizer, some fuel *and* food, instead of the normal food vs fuel debate. As the synfuel would be used on-site, the process would be more efficient and might just provide enough fuel for a new organic-agriculture fusion.

It has been suggested that some transgenic plants may some day be developed which would allow for maintaining or increasing yields while requiring fewer fossil-fuel-derived inputs than conventional crops. The possibility of success of these programs is questioned by ecologists and economists concerned with unsustainable GMO practices such as terminator seeds.

While there has been some research on sustainability using GMO crops, at least one prominent multi-year attempt by Monsanto Company has been unsuccessful, though during the same period traditional breeding techniques yielded a more sustainable variety of the same crop.

Electrical Energy Efficiency on Farms

Policy

Agricultural policy focuses on the goals and methods of agricultural production. At the policy level, common goals of agriculture include:

- Conservation
- Economic stability
- Environmental sustainability
- Food quality: Ensuring that the food supply is of a consistent and known quality.
- Food safety: Ensuring that the food supply is free of contamination.
- Food security: Ensuring that the food supply meets the population's needs.
- Poverty reduction

Agronomy

Agronomy is the science and technology of producing and using plants for food, fuel, feed, fibre, and reclamation. Agronomy encompasses work in the areas of plant genetics, plant physiology, meteorology, and soil science. Agronomy is the application of a combination of sciences like biology, chemistry, economics, ecology, earth science, and genetics. Agronomists today are involved with many issues including producing food, creating healthier food, managing environmental impact of agriculture, and creating energy from plants. Agronomists often specialise in areas such as crop rotation, irrigation and drainage, plant breeding, plant physiology, soil classification, soil fertility, weed control, insect and pest control.

Plant Breeding

This area of agronomy involves selective breeding of plants to produce the best crops under various conditions. Plant breeding has increased crop yields and has improved the nutritional value of numerous crops, including corn, soybeans, and wheat. It has also led to the development of new types of plants. For example, a hybrid grain called triticale was produced by crossbreeding rye and wheat. Triticale contains more usable protein than does either rye or wheat. Agronomy has also been instrumental in fruit and vegetable production research. It is understood that the role of agronomist includes seeing whether produce from a field of 'x' meets the following conditions: 1. Land and water access, 2. Commercialisation (market), 3. Quality and quantity of inputs, 4. Risk protection (insurance), 5. Agricultural credit.

Biotechnology

Agronomists use biotechnology to extend and expedite the development of desired characteristics listed in the Plant Breeding section. Biotechnology is often a lab activity requiring field testing of the new crop varieties that are developed. In addition to increasing crop yields agronomic biotechnology is increasingly being applied for novel uses other than food. For example, oilseed is at present used mainly for margarine and other food oils, but it can be modified to produce fatty acids for detergents, substitute fuels and petrochemicals.

Soil Science

Agronomists study sustainable ways to make soils more productive and profitable. They classify soils and reproduce them to determine whether they contain substances vital to plant growth such as compounds of nitrogen, phosphorus, and potassium. If a certain soil is deficient in these substances, fertilizers may provide them. Soil science also involves investigation of the movement of nutrients through the soil, the amount of nutrients absorbed by a plant's roots, and the development of roots and their relation to the soil.

Soil Conservation

In addition, agronomists develop methods to preserve the soil and to decrease the effects of erosion by wind and water. For example, a technique called contour plowing may be used to prevent soil erosion and conserve rainfall. Researchers in agronomy also seek ways to use the soil more effectively in solving other problems. Such problems include the disposal of human and animal wastes; water pollution; and the build-up in the soil of pesticides. No-tilling crops is a technique now used to help prevent erosion. Planting of soil binding grasses along contours can be tried in steep slopes. For better effect, contour drains of depths up to 1 metre may help retain the soil and prevent permanent wash off.

Agroecology

Agroecology is the management of agricultural systems with an emphasis on ecological and environmental perspectives. This area is closely associated with work in the areas of sustainable agriculture, organic farming, alternative food systems and the development of alternative cropping systems.

Theoretical Modelling

Agronomy Schools

Agronomy programs are offered at colleges, universities, and specialised agricultural schools. Agronomy programs often involve classes across a range of departments including agriculture, biology, chemistry, and physiology. They can usually take from four to twelve years. Many companies will pay an agronomist-in-training's way through college if they agree to work for them when they graduate.

Career Outlook

Due to the continued growth of the global population—and the consequent expanding need for study of food crops and agriculture in general—the outlook for agronomy and agronomists is excellent. Past agricultural research has created higher yielding crops, crops with better resistance to pests and plant pathogens, and more effective fertilizers and pesticides. Research is still necessary, however, particularly as insects and diseases continue to adapt to pesticides and as soil fertility and water quality continue to need improvement.

Emerging biotechnologies will play an ever larger role in agricultural research. Scientists will be needed to apply these technologies to the creation of new food products and other advances. Moreover, increasing demand is expected for biofuels and other agricultural products used in industrial processes. Agricultural scientists will be needed to find ways to increase the output of crops used in these products.

Agronomists will also be needed to balance increased agricultural output with protection and preservation of soil, water, and ecosystems. They increasingly encourage the practice of sustainable agriculture by developing and implementing plans to manage pests, crops, soil fertility and erosion, and animal waste in ways that reduce the use of harmful chemicals and do little damage to farms and the natural environment.

Most agronomists are consultants, researchers, or teachers. Many work for agricultural experiment stations, federal or state government agencies, industrial firms, or universities. Agronomists also serve in such international organisations as the Agency for International Development, The United States Department of Agriculture, and the Food and Agriculture Organisation of the United Nations.

Agronomists career options are expanding rapidly with possible ties with golf landscaping including topsoil analysis and drainage conditions. They often work in conjunction with landscape architects and engineers to determine the best soil qualities/conditions to suit the site specifications.

Chapter 7

Biodynamic Agriculture

Biodynamic agriculture is a method of organic farming that emphasizes the holistic development and interrelationships of the soil, plants and animals as a self-sustaining system. Biodynamic farming has much in common with other organic approaches, such as emphasizing the use of manures and composts and excluding of the use of artificial chemicals on soil and plants. There are independent certification agencies for biodynamic products; most of these agencies are members of the international biodynamics standards group Demeter International.

Biodynamics was one of the first modern ecological farming systems and is considered to be one of the most sustainable. As of 2011 biodynamic techniques were used on 142,482 hectares in 47 countries; Germany accounts for 45.1% of the global total. Methods unique to the biodynamic approach include the use of fermented herbal and mineral preparations as compost additives and field sprays and the use of an astronomical sowing and planting calendar. Biodynamics originated out of the work of Rudolf Steiner, the founder of anthroposophy.

Biodynamic agriculture has been characterised as pseudoscience.

History

The development of biodynamic agriculture began in 1924 with a series of eight lectures on agriculture given by philosopher Rudolf Steiner at Schloss Koberwitz in Silesia, Germany, (now Kobierzyce in Poland east of Wroc³aw). The course was held in response to a request by farmers who noticed degraded soil conditions and a deterioration in the health and quality of crops and livestock resulting from the use of chemical fertilizers. At the course of eight lectures there were 111 attendees coming from six countries. An agricultural research group was subsequently formed to test the effects of biodynamic methods on the life and health of soil, plants andanimals.

Today biodynamics is practiced in more than 50 countries worldwide. Demeter International is the primary certification agency for farms and gardens using the methods.

Geographic Developments

- Steiner's *Agriculture Course* was published in November 1924; an English translation appeared in 1928.

- In Australia the first biodynamic preparations were made by Ernesto Genoni in Melbourne in 1927 and by Bob Williams in Sydney in 1939. Since the 1950s research work has continued at the Biodynamic Research Institute (BDRI) in Powelltown, near Melbourne Australia under the direction of Alex Podolinsky. In 1989 Biodynamic Agriculture Australia was established, as a not for profit association.

- In 1938 Ehrenfried Pfeiffer's *Bio-Dynamic Farming and Gardening* was published in English, German, Dutch, French, and Italian.

- In the United States, the Biodynamic Farming & Gardening Association was founded in 1938 as a New York state corporation.

- In the UK in 1939, Britain's first biodynamic agriculture conference, the Betteshanger Summer School and Conference on Biodynamic Agriculture, was held at Lord Northbourne's farm in Kent; Ehrenfried Pfeiffer was the lead presenter.

- In France the International Federation of Organic Agriculture Movements (IFOAM) was formed in 1972 with five founding members, one of which was the Swedish Biodynamic Association.

- The University of Kassel had a Department of Biodynamic Agriculture from 2006 to March 2011.

- Germany, Italy and India are reported to be the leading countries in biodynamic agriculture based on biodynamic hectares.

Biodynamic Method of Farming

Biodynamic agriculturalists conceive of the farm as an organically self-contained entity with its own individuality, within which organisms are interdependent. "Emphasis is placed on the integration of crops and livestock, recycling of nutrients, maintenance of soil, and the health and well being of crops and animals; the farmer too is part of the whole." Cover crops, green manures and crop rotations are used extensively and the farms foster biodiversity. Biodynamic farms often have a cultural component and encourage local community. Some biodynamic farms use the Community Supported Agriculture model, which has connections with social threefolding.

Biodynamic Preparations

Steiner prescribed nine different preparations to aid fertilization which are the cornerstone of biodynamic agriculture, and described how these were to be prepared. Steiner believed that these preparations transferred supernatural terrestrial and cosmic "forces" into the soil. The prepared substances are numbered 500 through 508, where the first two are used for preparing fields whereas the latter seven are used for making compost. A long term trial (DOK experiment) evaluating the biodynamic farming system in comparison with organic and conventional farming systems, found that preparations have influence on soil structure and micro-organisms enhancing soil fertility and increasing biodiversity. Regarding compost development beyond accelerating the initial phase of composting, some positive effects have been noted:

- The field sprays contain substances that stimulate plant growth include cytokinins.
- Some improvement in nutrient content of compost.

Field Preparations

Field preparations, for stimulating humus formation:

- 500: (horn-manure) a humus mixture prepared by filling the horn of a cow with cow manure and burying it in the ground (40–60 cm below the surface) in the autumn. It is left to decompose during the winter and recovered for use the following spring.
- 501: Crushed powdered quartz prepared by stuffing it into a horn of a cow and buried into the ground in spring and taken out in autumn. It can be mixed with 500 but usually prepared on its own (mixture of 1 tablespoon of quartz powder to 250 litres of water) The mixture is sprayed under very low pressure over the crop during the wet season, in an attempt to prevent fungal diseases. It should be sprayed on an overcast day or early in the morning to prevent burning of the leaves.

Both 500 and 501 are used on fields by stirring about one teaspoon of the of a horn in 40–60 litres of water for an hour and whirling it in different directions every second minute. Although some biodynamic beliefs refer to buried quartz "fermenting", a 2004 review commented that it is unclear what this actually means, as rock does not ferment.

Compost Preparations

Compost preparations, used for preparing compost, employ herbs which are frequently used in medicinal remedies:

- 502: Yarrow blossoms (*Achillea millefolium*) are stuffed into urinary bladders from Red Deer (*Cervus elaphus*), placed in the sun during summer, buried in earth during winter and retrieved in the spring.
- 503: Chamomile blossoms (*Matricaria recutita*) are stuffed into small intestines from cattle buried in humus-rich earth in the autumn and retrieved in the spring.
- 504: Stinging nettle (*Urtica dioica*) plants in full bloom are stuffed together underground surrounded on all sides by peat for a year.
- 505: Oak bark (*Quercus robur*) is chopped in small pieces, placed inside the skull of a domesticated animal, surrounded by peat and buried in earth in a place where lots of rain water runs past.
- 506: Dandelion flowers (*Taraxacum officinale*) is stuffed into the peritoneum of cattle and buried in earth during winter and retrieved in the spring.
- 507: Valerian flowers (*Valeriana officinalis*) are extracted into water.
- 508: Horsetail (*Equisetum*)

One to three grams (a teaspoon) of each preparation is added to a dung heap by digging 50 cm deep holes with a distance of 2 metres from each other, except for the 507 preparation, which is stirred into 5 litres of water and sprayed over the entire compost surface. All preparations are thus used in homeopathic quantities. Each compost preparation is designed to guide a particular decomposition process in the composting mass. One study found that the oak bark preparation improved disease resistance in zucchini.

Astronomical Planting Calendar

The approach considers that there are astronomical influences on soil and plant development, specifying, for example, what phase of the moon is most appropriate for planting, cultivating or harvesting various kinds of crops. This aspect of biodynamics has been termed "astrological" in nature.

Seed Production

Biodynamic agriculture has focused on open pollination of seeds (permitting farmers to grow their own seed) and the development of locally adapted varieties. The seed stock is not controlled by large, multinational seed companies.

Trademark Protection of Term Biodynamic

The term *Biodynamic* is a trademark held by the Demeter association of biodynamic farmers for the purpose of maintaining production standards used both in farming and processing foodstuffs.(This is not a trademark held privately in New Zealand) The trademark is intended to protect both the consumer and the producers of biodynamic produce. Demeter International is an organisation of member countries; each country has its own Demeter organisation which is required to meet international production standards (but can also exceed them). The original Demeter organisation was founded in 1928; the U.S. Demeter Association was formed in the 1980s and certified its first farm in 1982. In France, Biodivincertifies biodynamic wine. In Egypt, SEKEM has created the Egyptian Biodynamic Association (EBDA), an association that provides training for farmers to become certified.

Studies of Efficacy

Studies have compared biodynamic farming methods to both other organic methods and to conventional methods. Most studies have found that biodynamic farms have soil quality significantly better than conventionally farmed soils but comparable to the soil quality achieved by other organic methods; the decisive factor is likely to be the use of compost. Studies of yields differ in their conclusions, and have found:

- A long-term study conducted at a commercial vineyard in California compared vineyard blocks treated with biodynamic preparations alongside those tended with general organic farming methods, to examine effects upon soil and crop quality. "No differences were found in soil quality" during the first six years of the study, and analyses of other indicators including the yield per vine, clusters per vine, cluster and berry weight also showed there were no differences. The study did find a statistically significant (p-value < 0.05) difference in the yield-to-pruning weight ratio, indicating an "ideal vine balance for producing high-quality wine grapes" for the biodynamically treated crop, but noted the control vines had been "slightly overcropped". In one particular year of the study the biodynamically treated wine grapes had significantly higher Brix and notably higher total phenols and anthocyanins. In conclusion, the study found that biodynamic preparations "may affect" the vine canopy and chemistry, but showed no effects on the soil and tissue nutrient parameters measured in the study.

- Decomposition was significantly faster in plots which received farmyard manure (FYM) treated with biodynamic preparations than in plots which received no FYM, FYM without preparations or FYM with an alternative preparation. "The application of completely prepared FYM led to significantly higher biomass and abundance of endogeic or anecic earthworms than in plots where non-prepared FYM was applied."

- A 21-year study by Mäder *et al* for the FiBL Institute in Switzerland compared the agronomic and ecological performance of biodynamic, organic and two conventional systems. The study found that nutrient input in the biodynamic and organic systems was 34 to 51% lower than in the conventional systems but crop yield was only 20% lower on average, indicating more efficient production. The total energy (for fuel, production of mineral fertilizer and pesticides, etc.) to produce a dry-matter unit of crop was 20 to 56% lower for the biodynamic and organic systems, and pesticide input was reduced by 97% (by 100% for the biodynamic system). In regard to soil aggregate stability, soil pH, humus formation, soil calcium, microbial biomass, and faunal biomass (earthworms and arthropods), the biodynamic system was superior even to the organic system, which in turn had superior results over the conventional systems. With the significant increase in microbial diversity in the biodynamic and organic systems, there was a significant associated decrease in metabolic quotient, indicating a greater ability to use organic material for plant growth.

The methodology of this study was criticized as faulty by two wine writers: philosophy professor Douglass Smith and criminology professor Jesús Barquín. They said the methodology yielded false conclusions. Smith and Barquín wrote that, to separate the mystical elements in biodynamics from standard organic techniques, no comparison should ever be made between conventional and biodynamic farming, only between biodynamic and organic farming. They noted that the Mäder paper's supplemental materials, found online but not in the published paper, described significant differences between the biodynamic and the organic methodology; for instance, unspecified chemicals were added to the organic farm's compost, possibly leading to its slightly poorer performance.

- Biodynamic preparations increase the soil organic matter content.
- A further study investigated whether biodynamic preparations had any effect on the yield and growth of lentil and wheat crops, weed populations and soil fertility in the short term. The study found that "in general, soils and crops treated with biodynamic preparations showed few differences from those not treated". Plots tended with biodynamically treated compost produced results for yield, crop quality and soil fertility that were the same to those tended with non-biodynamic composts and NPK fertilizers. Some alteration was observed in the nitrogenous chemistry of the soil and grain where biodynamic field sprays were applied, however the study did not ascribe or discern any biological significance to the difference. Among the variables considered by the study, some measured outcomes correlated with biodynamic field spray usage, including a higher per-unit biomass yield ratio for lentils and a lowering of carbon and crude protein in wheat grains. The study's conclusion remarked that "any additional short-term benefits from biodynamic preparations remain questionable."
- Minor effects for the field sprays on the carbon content and soil microbial fatty acid profile, but no effects for the compost preparations
- A 1993 study compared soil quality and financial performance of Biodynamic and conventional farms in New Zealand. The study reported that, "The Biodynamic farms proved in most enterprises to have soils of higher biological and physical quality: significantly greater in organic matter, content and microbial activity, more earthworms, better soil structure, lower bulk density, easier penetrability, and thicker topsoil." The biodynamic farms were just as financially viable on a per hectare basis. The study compared biodynamic farms with adjacent conventional farms, but didn't attempt to compare farms of similar size, or with similar crops.

Criticism

Biodynamic agriculture has been criticized as pseudoscience by scholars. In a 2002 newspaper editorial, Peter Treue, a researcher with the University of Kiel, characterised biodynamics as pseudoscience and argued that similar or equal results can be obtained using standard organic farming principles. He wrote that the biodynamic preparations more resemble alchemy or magic akin to geomancy.

In a 1994 analysis, Holger Kirchmann, a soil researcher with the Swedish University of Agricultural Sciences, concluded that Steiner's instructions were occult and dogmatic, and cannot contribute to the development of alternative or sustainable agriculture and that many of Steiner's statements are not provable because scientifically clear hypotheses cannot be made from his descriptions (for example, it is hard to prove that one has harnessed "cosmic forces" in the foods). Kirchmann asserted that when methods of biodynamic agriculture were tested scientifically, the results were unconvincing.

Further, in a 2004 overview of biodynamic agriculture, Linda Chalker-Scott, a researcher at Washington State University, characterised biodynamics as pseudoscience, writing that Steiner did not use scientific methods to formulate his theory of biodynamics, and that the later addition of valid organic farming techniques has "muddled the discussion" of Steiner's original idea.

Based on the scant scientific testing of biodynamics, Chalker-Scott concluded "no evidence exists" that homeopathic preparations improve the soil. In Michael Shermer's two-volume work, *The Skeptic Encyclopedia of Pseudoscience*, biodynamic agriculture is described as relying on astrological conditions, cosmic influences and magical rituals.

Biological Pest Control

Biological control is defined as the reduction of pest populations by natural enemies and typically involves an active human role. Natural enemies of insect pests, also known as biological log control agents; and include predators, parasitoids, and pathogens. Biological control agents of plant diseases are most often referred to as antagonists.

Biological control agents of weeds include herbivores and plant pathogens. Predators, such as birds, lady beetles and lacewings, are mainly free-living species that consume a large number of prey during their whole lifetime. Parasitoids are species whose immature develops on or within a single insect host, ultimately killing or fatally infecting the host. Most have a very narrow host range. Many species of wasps and some flies are parasitoids. Pathogens are disease-causing organisms including bacteria, fungi, and viruses. They kill or debilitate their own host and are relatively specific. There are three basic types of biological control strategies; conservation, classical biological control, and augmentation.

Conservation

The conservation of natural enemies is probably the most important and readily available biological control practice available to homeowners and gardeners. Natural enemies occur in all areas, from the private garden to the open field. They are adapted to the habitat and to the target pest, and their conservation is generally simple and cost-effective. Lacewings, lady beetles, hover fly larvae, and parasitized aphid mummies are almost always present in aphid colonies. Fungus-infected adult flies are often common following periods of high humidity. These naturally occurring biological controls are often susceptible to the same pesticides used to target their pest hosts. Preventing the accidental eradication of natural enemies is termed "simple conservation."

To conserve and encourage pest insect eating birds, including native plants and ornamental plants that supply berries, acorns, nuts, seeds, nectar, and other vegetative foods, and also bird nest building materials, encourages their presence, health, and new generations. These qualities can also increase the visible population to enjoy in a garden. Using companion planting and the birds' insect cuisine habits is a traditional method for biological control agent pest control in anorganic garden and any landscape, and in organic farming and sustainable agriculture. Installing specified nest boxes for mosquito-eating bats reduces a pest and increases endangered species conservation.

Effects of Biological Control

Effects on Native Biodiversity

The cane toad, Bufo marinus

Biological control can potentially have positive and negative effects on biodiversity. Usually a biological control is introduced to an area to protect a native species from an invasive or exotic species that has moved into its area. The control is introduced to lessen the competition between native and introduced species. However, the introduced control does not always target only the intended species. It can also target native species.

When introducing a high biological control to a new area, a primary concern is the host- or prey-specificity of the control agent. Generalist feeders (control agents that are not restricted to a single species or a small range of species) often make poor biological control agents, and may become invasive species themselves. For this reason, potential biological control agents should be subject to extensive testing and quarantine before release into any new environment. If a species is introduced and attacks a native species, the biodiversity in that area can change dramatically. When one native species is removed from an area, it may have filled an essential ecological niche. When this niche is absent it may directly affect the entire ecosystem.

Because they tend to be generalist feeders, vertebrate animals seldom make good biological control agents, and many of the classic cases of "biocontrol gone awry" involve vertebrates. For example, the cane toad, *Bufo marinus*, was introduced as a biological control and had significant negative impact on biodiversity. The cane toad was intentionally introduced to Australia to control the introduced French's Cane Beetle and the Greyback Cane Beetle. When introduced, the cane toad thrived very well and did not only feed on cane beetles but other insects as well. The cane toad soon spread very rapidly, thus taking over native amphibian habitat. The introduction of the cane toad also brought foreign disease to native reptiles. This drastically reduced the population of native toads and frogs. "The cane toad also exudes and can squirt poison from the parotid glands on their shoulders when threatened or handled. This toxin contains a cocktail of chemicals that can kill animals that eat it. Freshwater crocodiles, goannas, tiger snakes, dingos and northern quolls have all died after eating cane toads, as have pet dogs (Cane toad, 2003). This example shows how a small mis-introduced organism can alter the native biodiversity in a large ecosystem in a very expedient manner. A pyramid effect can take place if native species are reduced or eradicated. The domino effect keeps on going and can potentially exude on other bordering ecosystems until an equilibrium is reached.

A second example of a biological control agent that subsequently crossed over to native species is the *Rhinocyllus conicus*. The seed feeding weevil was introduced to North America to control exotic thistles (Musk and Canadian). However, the weevil did not target only the exotic thistles, it also targeted native thistles that are essential to various native insects. The native insects rely solely on native thistles and do not adapt to other plant species. Therefore, they cannot survive. Biological controls do not always have negative impacts on biodiversity (Corry 2000).

Successful biological control reduces the population density of the target species over several years, thus providing the potential for native species to re-establish. In addition, regeneration and re-establishment programs can aid to the recovery of native species. Native species can be affected in a positive way as well. To develop or find a biological control that exerts control only on the targeted species is a very lengthy process of research and experiments. In the late 19th century, the citrus industry was in great fear when the cottony cushion scale was discovered. This organism could cause a great deal of economic loss to the industry. However, a biological control was introduced. The vedalia beetle and a parasitoidfly were introduced to control the pest. Within a few years time, the cottony cushion scale was controlled by the natural enemies and the citrus industry suffered little financial loss. Many exotic or invasive species can suppress the development of native species. The introduction of an effective biological control that reduces the population of the invasive species allows the rejuvenation of the native species. Biological controls can reduce competition for biotic and a bioticfactors which can result in the re-establishment of the once over ran native species.

Effects on Invasive Species

The invasive species Alternanthera philoxeroides (alligator weed) was controlled in Florida (U.S.) by the introduction of Agasicles hygrophila (alligator weed flea beetle)

Invasive species are closely associated with biological controls because the environment in which they are invasive most likely does not contain their natural enemies. If invasive species are not controlled, biodiversity may be at great threat in the affected area. An example of an invasive species is the alligator weed. This plant was introduced to the United States from South America. This aquatic weed spreads rapidly and causes many problems in lakes and rivers. The weed takes root in shallow water causing major problems for navigation, irrigation, and flood control. The alligator weed flea beetle and two other biological controls were released in Florida. Because of their success, Florida banned the use of herbicides to control alligator weed three years after the controls were introduced. Similarly, *Galerucella calmariensis*, a leaf beetle, has been introduced in North America as a control agent for purple loosestrife (*Lythrum salicaria*).

Biological controls for invasive species also can have a negative impact on biodiversity. The cane toad, as mentioned previously, is an example of trying to control an invasive species. The cane toad was introduced to eradicate an invasive species. It became invasive, thus altering the biodiversity. The introduction of the cane toad could have caused more of a disturbance in biodiversity than the targeted species did.

Classical Biological Control

Classical biological control is the introduction of natural enemies to a new locale where they did not originate or do not occur naturally. This is usually done by government authorities. In many instances the complex of natural enemies associated with an insect pest may be inadequate. This is especially evident when an insect pest is accidentally introduced into a new geographic area without its associated natural enemies. These introduced pests are referred to as exotic pests and comprise about 40% of the insect pests in the United States. Examples of introduced vegetable pests include the European corn borer (*Ostrinia nubilalis*), one of the most destructive insects in North America. To obtain the needed natural enemies, scientists turned to classical biological control. This is the practice of importing, and releasing for establishment, natural enemies to control an introduced (exotic) pest, although it is also practiced against native insect pests. The first step in the process is to determine the origin of the introduced pest and then collect appropriate natural enemies associated with the pest or closely related species. The natural enemy is then passed through a rigorous quarantine process, to ensure that no unwanted organisms (such as hyperparasitoids) are introduced, then they are mass produced, and released. Follow-up studies are conducted to determine if the natural enemy becomes successfully established at the site of release, and to assess the long-term benefit of its presence.

There are many examples of successful classical biological control programs.

Joseph Needham noted a Chinese text dating from 304AD, *Records of the Plants and Trees of the Southern Regions*, by Hsi Han, which describes mandarin oranges protected by biological pest control techniques that are still in use today.

One of the earliest successes in the west was in controlling *Icerya purchasi*, the cottony cushion scale, a pest that was devastating the California citrus industry in the late 19th century. A predatory insect *Rodolia cardinalis* (the Vedalia Beetle), and a parasitoid fly were introduced from Australia by Charles Valentine Riley. Within a few years the cottony cushion scale was completely controlled by these introduced natural enemies.

Damage from *Hypera postica* Gyllenhal, the alfalfa weevil, a serious introduced pest of forage, was substantially reduced by the introduction of several natural enemies. 20 years after their introduction the population of weevils in the alfalfa area treated for alfalfa weevil in the Northeastern United States was reduced by 75 percent. A small wasp, *Trichogramma ostriniae*, was introduced from China to help control the European corn borer making it a recent example of a long history of classical biological control efforts for this major pest. Many classical biological control programs for insect pests and weeds are under way across the United States and Canada. The population of *Levuana irridescens* (the Levuana moth), a serious coconut pest in Fiji, was brought under control by a classical biological control program in the 1920s.

Classical biological control is long lasting and inexpensive. Other than the initial costs of collection, importation, and rearing, little expense is incurred. When a natural enemy is successfully established it rarely requires additional input and it continues to kill the pest with no direct help from humans and at no cost. Classical biological control does not always work. It is usually most effective against exotic pests and less so against native insect pests. The reasons for failure are not often known but may include the release of too few individuals, poor adaptation of the natural enemy to environmental conditions at the release location, and lack of synchrony between the life cycle of the natural enemy and host pest.

Augmentation

This third type of biological control involves the supplemental release of natural enemies. Relatively few natural enemies may be released at a critical time of the season (inoculative release) or literally millions may be released (inundative release). Additionally, the cropping system may be modified to favour or augment the natural enemies. This latter practice is frequently referred to as habitat manipulation. An example of inoculative release occurs in greenhouse production of several crops. Periodic releases of the parasitoid, *Encarsia formosa*, are used to control greenhouse whitefly, and the predaceous mite, *Phytoseiulus persimilis*, is used for control of the two-spotted spider mite. Lady beetles, lacewings, or parasitoids such as those from the genus *Trichogramma* are frequently released in large numbers (inundative release). Recommended release rates for Trichogramma in vegetable or field crops range from 5,000 to 200,000 per acre (1 to 50 per square metre) per week depending on level of pest infestation. Similarly, entomopathogenic nematodes are released at rates of millions and even billions per acre for control of certain soil-dwelling insect pests.

A turnaround flowerpot, filled with straw to attract Dermaptera-species

The spraying of octopamine analogues (such as 3-FMC) has been suggested as a way to boost the effectiveness of augmentation.

Octopamine, regarded as the invertebrate counterpart of dopamine plays a role in activating the insects' flight-or-fight response. The idea behind using octopamine analogues to augment biological control is that natural enemies will be more effective in their eradication of the pest, since the pest will be behaving in an unnatural way because its flight-or-fight mechanism has been activated. Octopamine analogues are purported to have two desirable characteristics for this type of application: (1) they affect insects at very low dosages (2) they do not have a physiological effect in humans (or other vertebrates).

Habitat or environmental manipulation is another form of augmentation. This tactic involves altering the cropping system to augment or enhance the effectiveness of a natural enemy. Many adult parasitoids and predators benefit from sources of nectar and the protection provided by refuges such as hedgerows, cover crops, and weedy borders. Also, the provisioning of natural shelters in the form of wooden caskets, boxes or (turnaround) flowerpots is a form of this. For example, the stimulation of the natural predator *Dermaptera* is done in gardens by hanging up turnaround flowerpots with straw or wood wool.

Mixed plantings and the provision of flowering borders can increase the diversity of habitats and provide shelter and alternative food sources. They are easily incorporated into home gardens and even small-scale commercial plantings, but are more difficult to accommodate in large-scale crop production. There may also be some conflict with pest control for the large producer because of the difficulty of targeting the pest species and the use of refuges by the pest insects as well as natural enemies.

Examples of habitat manipulation include growing flowering plants (pollen and nectar sources) such as Buckwheat near crops to attract and maintain populations of natural enemies. For example, hover fly adults can be attracted to umbelliferous plants in bloom.

Biological control experts in California have demonstrated that planting prune trees in grape vineyards provides an improved overwintering habitat or refuge for a key grape pest parasitoid. The prune trees harbour an alternate host for the parasitoid, which could previously overwinter only at great distances from most vineyards. Caution should be used with this tactic because some plants attractive to natural enemies may also be hosts for certain plant diseases, especially plant viruses that could be vectored by insect pests to the crop. Although the tactic appears to hold much promise, only a few examples have been adequately researched and developed.

Examples of Predators

Lacewings are available from biocontrol dealers.

Ladybugs, and in particular their larvae which are active between May and July in the northern hemisphere, are voracious predators of aphids such as greenfly and blackfly, and will also consumemites, scale insects and small caterpillars. The ladybug is a very familiar beetle with various coloured markings, whilst its larvae are initially small and spidery, growing up to 17 mm long. The larvae have a tapering segmented grey/black body with orange/yellow markings and ferocious mouthparts. They can be encouraged by cultivating a patch of nettles in the garden and by leaving hollow stems and some plant debris over winter so that they can hibernate.

Predatory Polistes wasp looking for bollworms or other caterpillars on a cotton plant

Hoverflies resemble slightly darker bees or wasps and they have characteristic hovering, darting flight patterns. There are over 100 species of hoverfly whose larvae principally feed upon greenfly, one larva devouring up to fifty a day, or 1000 in its lifetime. They also eat fruit tree spider mites and small caterpillars. Adults feed on nectar and pollen, which they require for egg production. Eggs are minute (1 mm), pale yellow white and laid singly near greenfly colonies. Larvae are 8–17 mm long, disguised to resemble bird droppings, they are legless and have no distinct head. Semi-transparent in a range of colours from green, white, brown and black.

Hoverflies can be encouraged by growing attractant flowers such as the poached egg plant *(Limnanthes douglasii)*, marigolds or phacelia throughout the growing season.

Dragonflies are important predators of mosquitoes, both in the water, where the dragonfly naiads eat mosquito larvae, and in the air, where adult dragonflies capture and eat adult mosquitoes. Community-wide mosquito control programs that spray adult mosquitoes also kill dragonflies, thus removing an important biocontrol agent, and can actually increase mosquito populations in the long term.

Other useful garden predators include lacewings, pirate bugs, rove and ground beetles, aphid midge, centipedes, spiders, predatory mites, as well as larger fauna such as frogs, toads, lizards, hedgehogs, slow-worms and birds. Cats and rat terriers kill field mice, rats, June bugs, and birds. Dogs chase away many types of pest animals. Dachshunds are bred specifically to fit inside tunnels underground to kill badgers.

More examples:

- *Phytoseiulus persimilis* (against spider mites)
- *Amblyseius californicus* (against spider mites)
- *Amblyseius cucumeris* (against spider mites)
- *Typhlodromips swirskii* (against spider mites, thrips, and white flies)
- *Feltiella acarisuga* (against spider mites)
- *Stethorus punctillum* (against spider mites)
- *Macrolophus caluginosus* (against spider mites).

Parasitoid Insects

Most insect parasitoids are wasps or flies. Parasitoids comprise a diverse range of insects that lay their egg on or in the body of an insect host, which is then used as a food for developing larvae. Parasitic wasps take much longer than predators to consume their victims, for if the larvae were to eat too fast they would run out of food before they became adults. Such parasites are very useful in the organic garden, for they are very efficient hunters, always at work searching for pest invaders. As adults they require high energy fuel as they fly from place to place, and feed upon nectar, pollen and sap, thereby pollinating plenty of flowering plants, particularly buckwheat, umbellifers, and composites will encourage their presence.

Four of the most important groups are:

- Ichneumonid wasps: (5–10 mm). Prey mainly on caterpillars of butterflies and moths.
- Braconid wasps: Tiny wasps (up to 5 mm) attack caterpillars and a wide range of other insects including greenfly. A common parasite of the cabbage white caterpillar- seen as clusters of sulphur yellow cocoons bursting from collapsed caterpillar skin.
- Chalcid wasps: Among the smallest of insects (<3 mm). Parasitize eggs/larvae of greenfly, whitefly, cabbage caterpillars, scale insects and Strawberry Tortrix Moth (*Acleris comariana*).
- Tachinid flies: Parasitize a wide range of insects including caterpillars, adult and larval beetles, true bugs, and others.

Examples of parasitoids: wasp:

- *Encarsia formosa* (against white flies)
- *Eretmocerus* spp. (against white flies)
- *Aphidius colemani* (against aphids).

Biological Control with Micro-organisms

Various microbial insect diseases occur naturally, but may also be used as biological pesticides. When naturally occurring, these outbreaks are density-dependent in that they generally only occur as insect populations become denser.

Bacteria and Biological Control

Bacteria used for biological control infect insects via their digestive tracts, so insects with sucking mouth parts like aphids and scale insects are difficult to control with bacterial biological control. *Bacillus thuringiensis* is the most widely applied species of bacteria used for biological control, with at least four sub-species used to control Lepidopteran (moth, butterfly), Coleopteran (beetle) and Dipteran (true flies) insect pests.

Fungi and Biological Control

Fungi that cause disease in insects are known as entomopathogenic fungi, including at least fourteen species of entomophthoraceous fungi attack aphids. Species in the genus *Trichoderma* are used to manage some soilborne plant pathogens. *Beauveria bassiana* is used to manage different types of pest such whiteflies, thrips, aphids and weevils.

Examples of entomopathogenic fungi:

- *Beauveria bassiana* (against white flies, thrips, aphids and weevils)
- *Paecilomyces fumosoroseus* (against white flies, thrips and aphids)
- *Metarhizium* spp. (against beetles, locusts, Hemiptera, spider mites and other pests)
- *Lecanicillium lecanii* (against white flies, thrips and aphids)
- *Cordyceps* species (sometines teleomorphs of the above: that infect a wide spectrum of arthropods).

Combined use of Parasitoids and Pathogens

In cases of massive and severe infection of invasive pests, techniques of pest control are often used in combination. An example being, that of the emerald ash borer (*Agrilus planipennis* Fairmaire, family Buprestidae), an invasive beetle from China, which has destroyed tens of millions of ash trees in its introduced range in North America's part of the campaign against the emerald ash borer (EAB), American scientists in conjunction with the Chinese Academy of Forestry searched since 2003 for its natural enemies in the wild leading to the discovery of several parasitoid wasps, namely *Tetrastichus planipennisi*, a gregarious larval endoparasitoid, *Oobius agrili*, a solitary, parthenogenic egg parasitoid, and *Spathius agrili*, a gregarious larval ectoparasitoid.

These have been introduced and released into the United States of America as a possible biological control of the emerald ash borer. Initial results have shown promise with *Tetrastichus planipennisi* and it is now being released along with *Beauveria bassiana*, a fungal pathogen with known insecticidal properties.

Plants to Regulate Insect Pests

Choosing a diverse range of plants for the garden can help to regulate pests in a variety of ways, including;

- Masking the crop plants from pests, depending on the proximity of the companion or intercrop.
- Producing olfactory inhibitors, odours that confuse and deter pests.

- Acting as trap plants by providing an alluring food that entices pests away from crops.

- Serving as nursery plants, providing breeding grounds for beneficial insects.

- Providing an alternative habitat, usually in a form of a shelterbelt, hedgerow, or beetle bank where beneficial insects can live and reproduce. Nectar-rich plants that bloom for long periods are especially good, as many beneficials are nectivorous during the adult stage, but parasitic or predatory as larvae. A good example of this is the soldier beetle which is frequently found on flowers as an adult, but whose larvae eat aphids, caterpillars, grasshopper eggs, and other beetles.

- Some plants have chemical defences in order to regulate pests. The geranium has developed such a defence against Japanese beetles, one of the most damaging and expensive pests to control when it comes to ornamental and turf plants. The geranium's petals contain a chemical compound that paralyses the beetle within 30 minutes of ingestion. The beetle will remain paralysed for several hours and will typically regain movement within 24 hours. However, while paralysed the beetle is very vulnerable to its predators and is usually hunted before the paralysis subsides. Agricultural Research Service (ARS) scientists are working to isolate the chemical compound in geraniums that causes the paralysis in the beetles. Scientists hope to one day use this natural pesticide to control the population of beetles. In addition to this research ARS scientists are studying ways to help geranium leaves better hold on to pesticide chemicals that are sprayed on them, that way less pesticides will have to be applied to the leaves.

Plants to Regulate Plants

The legume vine *Mucuna pruriens* is used in the countries of Benin and Vietnam as a biological control for problematic *Imperata cylindrica* grass. *Mucuna pruriens* is said not to be invasive outside its cultivated area. *Desmodium uncinatum*can be used in push-pull farming to stop the parasitic plant, *Striga*.

Directly Introducing Biological Controls

Most of the biological controls listed above depend on providing incentives in order to 'naturally' attract beneficial insects to the garden. However there are occasions when biological controls can be directly introduced. Common biocontrol agents include parasitoids, predators, pathogens or weed feeders. This is particularly appropriate in situations such as the greenhouse, a largely artificial environment, and are usually purchased by mail order.

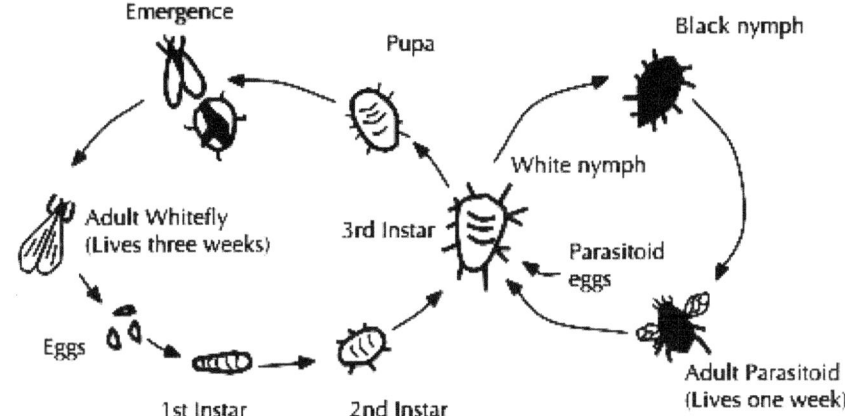

Diagram illustrating the life cycles of Greenhouse whitefly and its parasitoid wasp Encarsia formosa

Some biocontrol agents that can be introduced include;

- *Encarsia formosa.* This is a small predatory chalcid wasp which is parasitical on whitefly, a sap-feeding insect which can cause wilting and black sooty moulds. It is most effective when dealing with low level infestations, giving protection over a long period of time. The wasp lays its eggs in young whitefly 'scales', turning them black as the parasite larvae pupates. It should be introduced as soon as possible after the first adult whitefly are seen. Should be used in conjunction with insecticidal soap.

- Red spider mite, another pest found in the greenhouse, can be controlled with the predatory mite *Phytoseilus persimilis.* This is slightly larger than its prey and has an orange body. It develops from egg to adult twice as fast as the red spider mite and once established quickly overcomes infestation.

- A fairly recent development in the control of slugs is the introduction of 'Nemaslug', a microscopic nematode (*Phasmarhabditis hermaphrodita*) which will seek out and parasitize slugs, reproducing inside them and killing them. The nematode is applied by watering onto moist soil, and gives protection for up to six weeks in optimum conditions, Nemaslug nematodes are mainly effective with small and young slugs under the soil surface.

- A bacterial biological control which can be introduced in order to control butterfly caterpillars is *Bacillus thuringiensis*. This available in sachets of dried spores which are mixed with water and sprayed onto vulnerable plants such as brassicas and fruit trees. The bacterial disease will kill the caterpillars, but leave other insects unharmed. There are strains of *Bt* that are effective against other insect larvae. *Bt israelensis* is effective against mosquito larvae and some midges.

The European Rabbit (Oryctolagus cuniculus) is seen as a major pest in Australia and New Zealand.

- A viral biological control which can be introduced in order to control the overpopulation of European rabbit in Australia is the rabbit haemorrhagic disease virus that causes the rabbit haemorrhagic disease.
- A biological control being developed for use in the treatment of plant disease is the fungus *Trichoderma viride*. This has been used against Dutch Elm disease, and to treat the spread of fungal and bacterial growth on tree wounds. It may also have potential as a means of combating silver leaf disease.
- Several species of dung beetle were introduced to Australia from South Africa and Europe during the Australian Dung

Beetle Project (1965–1985) led by Dr. George Bornemissza of the Commonwealth Scientific and Industrial Research Organisation in order to biologically control the population of pestilent bush flies and parasitic worms.

- The parasitoid *Gonatocerus ashmeadi* (Hymenoptera: Mymaridae) has been introduced to control the glassy-winged sharpshooter *Homalodisca vitripennis* (Hemipterae: Cicadellidae) in French Polynesia and has successfully controlled ~95% of the pest density.

Negative Consequences of Biological Pest Control

In some cases, biological pest control could have unforeseen averse consequences that outweigh all benefits, often by becoming an invasive species. For example:

- When the mongoose was introduced to Hawaii in order to control the rat population, it preyed on the endemic birds of Hawaii, especially their eggs, more often than it ate the rats. (Note, however, that the introduction of the mongoose was not undertaken based on scientific—or perhaps any— understanding of the consequences of such an action. The introduction of a generalist mammal for biocontrol of anything would be unthinkable by any reasonable standards today.)

- Cane toads (*Bufo marinus*) were introduced to Australia in the 1930s in a failed attempt to control the cane beetle, a pest of sugar cane crops. 102 toads were obtained from Hawaii and bred in captivity to increase their numbers until they were released into the sugar cane fields of the tropic north in 1935. It was later discovered that the toads could not jump very high and so they could not eat the cane beetles which stayed up on the upper stalks of the cane plants. The toad population grew dramatically and eventually out-competed native species. Not only were these toads very harmful to the Australian environment, they were also very toxic to would-be predators such as the native snakes.

- 5 cats brought to the subantarctic Marion Islands to catch mice in 1949 multiplied to 3,400 in about two decades and started to threaten local extinction of birds. They had to be infected with feline distemper virus. The rest were shot and completely eliminated by the 1990s.

- The sturdy and prolific mosquito fish was introduced from around the Gulf of Mexico to around the world in the 1930s and 40s to combat malaria; however, it was found to cause the decline of local fish and frogs through competition of other food source as well as eating their eggs.

Living organisms, through the process of evolution, may achieve increased resistance to biological, chemical, and physical methods of control over time. In the event the target pest population is not completely exterminated or is still capable of reproduction (were the pest control means a form of sterilization), the surviving population could acquire a tolerance to the applied pressures- this can result in an evolutionary arms race with the control method.

Community-supported Agriculture

Community-supported agriculture, a form of an alternative food network, (in Canada Community Shared Agriculture) (CSA) is a socio-economic model of agriculture and food distribution. A CSA consists of a community of individuals who pledge support to a farming operation where the growers and consumers share the risks and benefits of food production. CSAs usually consist of a system of weekly delivery or pick-up of vegetables and fruit, in a vegetable box scheme, and sometimes includes dairy products and meat.

History

Community-supported agriculture began in the early 1960s in Germany, Switzerland and Japan as a response to concerns about food safety and the urbanisation of agricultural land. In the 1960s groups of consumers and farmers in Europe formed cooperative partnerships to fund farming and pay the full costs of ecologically sound and socially equitable agriculture. In Europe, many of the CSA style farms were inspired by the economic ideas of Rudolf Steiner and experiments with community agriculture took place on farms using biodynamic agriculture. In 1965 mothers in Japan who were concerned about the rise of imported food and the loss of arable land started the first CSA projects called *Teikei* (Ðc:d) in Japanese – most likely unrelated to the developments in Europe.

The idea started to take root in the United States in 1984 when Jan VanderTuin brought the concept of CSA to North America from Europe. At the same time German Biodynamic farmer Trauger Groh and colleagues founded the Temple-Wilton Community Farm in Wilton, New Hampshire. VanderTuin had co-founded a community-supported agricultural project named Topinambur located near Zurich, Switzerland. Coinage of the term "community-supported agriculture" stems from Vander Tuin and the Great Barrington CSA that he co-founded with its proprietor Robyn Van En. Since that time community supported farms have been organised throughout North America — mainly in the Northwest, the Pacific coast, the Upper-Midwest and Canada. North America now has at least 13,000 CSA farms of which 12,549 are in the US according to the US Department of Agriculture in 2007. Some examples of larger and well established CSAs in the US are Angelic Organics and Roxbury Farm. CSA's have even become popular in urban environments as proven by the New York City Coalition Against Hunger's own CSA program that maintains locations in all five boroughs of the city. The largest subscription CSA with over 13,000 families is Farm Fresh To You in Capay Valley, California.

The CSA System

CSAs generally focus on the production of high quality foods for a local community, often using organic or biodynamic farming methods, and a shared risk membership–marketing structure. This kind of farming operates with a much greater degree of involvement of consumers and other stakeholders than usual — resulting in a stronger consumer-producer relationship. The core design includes developing a cohesive consumer group that is willing to fund a whole season's budget in order to get quality foods. The system has many variations on how the farm budget is supported by the consumers and how the producers then deliver the foods. CSA theory purports that the more a farm embraces whole-farm, whole-budget support, the more it can focus on quality and reduce the risk of food waste or financial loss.

Structure

In its most formal and structured European and North American forms CSAs focus on having:

- a transparent, whole season budget for producing a specified wide array of products for a set number of weeks a year;
- a common-pricing system where producers and consumers discuss and democratically agree to pricing based on the acceptance of the budget; and
- a 'shared risk and reward' agreement, *i.e.* that the consumers receive what the farmers grow even with the vagaries of seasonal growing.

Meaning that individuals, families, &/or groups do not directly pay for x pounds or kilograms of produce but rather support the budget of the whole farm and receive weekly what is seasonally ripe.

This approach eliminates the marketing risks, costs for the producer and an enormous amount of time and labour, and allows producers to focus on quality care of the soils, crops, animals and co-workers as well as on serving the customers. There is financial stability in this system which allows for thorough planning on the part of the farmer.

Some farms are dedicated entirely to their CSA while others also sell through on-farm stands, farmers' markets and other channels. Most CSAs are owned by the farmers while some offer shares in the farm as well as the harvest. Consumers have organised their own CSA projects and have gone as far as leasing land and hiring farmers. Many CSAs have a core group of members that assist with CSA administration. Some require or offer the option of members providing labour as part of the share price.

Some CSAs have evolved into social enterprises employing a number of local staff, improving the lot of local farmers and educating the local community about organic and ecologically responsible farming.

Typically CSA farms are small, independent, labour-intensive family farms. By providing a guaranteed market through prepaid annual sales consumers essentially help finance farming operations. This allows farmers to not only focus on quality growing but can also level the playing field in a food market that favours large-scale, industrialised agriculture over local food.

Vegetables and fruit are the most common CSA crops. Many CSAs practice ecological, organic or biodynamic agriculture by avoiding pesticides and inorganic fertilizers. The cost of a share is usually competitively priced when compared to the same amount of vegetables conventionally grown – partly because the cost of distribution is lowered.

Distribution and Marketing Methods

A distinctive feature of CSAs is the method of distribution. In the U.S. and Canada shares are usually provided weekly with pick-ups or deliveries occurring on a designated day and time. CSA subscribers often live in towns and cities – local drop-off locations, convenient to a number of members, are organised, often at the homes of members. Shares are also usually available on-farm.

CSAs are different from buying clubs and home delivery services where the consumer buys a specific product at a predetermined price. CSA members purchase only what the farm is able to successfully grow and harvest sharing some of the growing risk with the farmer. If the strawberry crop is not successful, for example, the CSA member will share the burden of the crop failure by receiving fewer, or lower quality, strawberries for the season. CSA members are often more actively involved in the growing and distribution process through shared newsletters and recipes, farm visits, farm workdays, advance purchases of shares and picking up their shares of produce.

Some families have enrolled in subscription CSAs in which a family pays a fixed price for each delivery and can start or stop the service as they wish. This kind of arrangement is also referred to as crop-sharing or box schemes. In such cases the farmer may supplement each box with produce brought in from neighbouring farms for a wider variety. Thus there is a distinction between the farmers selling prepaid shares in the upcoming season's harvest or a weekly subscription that represents that week's harvest. In all cases participants purchase a portion of the farm's harvest either by the season or by the week in return for what the farm is able to successfully grow and harvest.

An advantage of the close consumer-producer relationship is increased freshness of the produce because it does not have to be shipped long distances. The close proximity of the farm to the members also helps the environment by reducing pollution caused by transporting the produce. CSAs often include recipes and farm news in each box in which tours of the farm and work days are announced. Over a period of time consumers get to know who is producing their food and what production methods are used.

Share prices can vary dramatically depending on location. Variables also include the length of share season and average quantity and selection of food per share. As a rough average, in North America, a basic share may be $350–550 for a season lasting for 14–20 weeks in June to September (or October). The produce would be enough of each included crop for at least two people consisting of perhaps 8–12 common garden vegetables. Seasonal eating is implied as shares are usually based on the outdoor growing season which means a smaller selection at the beginning, and perhaps the end, of the period as well as a changing variety as the season progresses. Some CSA programs offer different share sizes or choices of share periods e.g. full-season and peak season. The film *The Real Dirt on Farmer John* documents the resurrection of a family farm through its conversion to a CSA model.

Similar Experiences Worldwide

The term CSA is mostly used in the USA but a variety of similar production and economic subsystems are in use worldwide:

- *Association pour le maintien de l'agriculture paysanne* (AMAP) in France,

- *Agriculture soutenue par la communauté* (ASC) in Québec,
- *Teikei* (Đc:d) in Japan,
- *Reciproco* in Portugal,
- *Landwirtschaftsgemeinschaftshof* in Germany,
- *Andelslandbruk* in Norway,
- *Gruppi di Acquisto Solidale* (GAS) in Italy, (, Ethical purchasing groups)

Solidarity Gardens

Orti Solidali (meaning Solidarity Gardens) is an example of a CSA in Italy. Reasons for participating are mostly on an ethical basis. The strong commitment of participants protected the development of the network from the market until properly established. Ethical values, cognitive frames, relational codes inside the network shape and constitute this protective environment. Orti Solidali uses a sustainable agronomic method for her a food production and creates goods while providing living wages and fair working conditions to the producers. With the aim of reduction of economic growth, also known as degrowth, they transition to a new economic system based on environmental protection and social equity.

Industrial Agriculture

Industrial farming is a form of modern farming that refers to the industrialised production of livestock, poultry, fish, and crops. The methods of industrial agriculture are technoscientific, economic, and political. They include innovation in agricultural machinery and farming methods, genetic technology, techniques for achieving economies of scale in production, the creation of new markets for consumption, the application of patent protection to genetic information, and global trade. These methods are widespread in developed nations and increasingly prevalent worldwide. Most of the meat, dairy, eggs, fruits, and vegetables available in supermarkets are produced using these methods of industrial agriculture.

Historical Development and Future Prospects

The birth of industrial agriculture more or less coincides with that of the Industrial Revolution in general. The identification of nitrogen, potassium, and phosphorus (referred to by the acronym NPK) as critical factors in plant growth led to the manufacture of synthetic fertilizers, making possible more intensive types of agriculture. The discovery of vitamins and their role in animal nutrition, in the first two decades of the 20th century, led to vitamin supplements, which in the 1920s allowed certain livestock to be raised indoors, reducing their exposure to adverse natural elements. The discovery of antibiotics and vaccines facilitated raising livestock in concentrated, controlled animal feed operations by reducing diseases caused by crowding. Chemicals developed for use in World War II gave rise to synthetic pesticides. Developments in shipping networks and technology have made long-distance distribution of agricultural produce feasible.

Agricultural production across the world doubled four times between 1820 and 1975 to feed a global population of one billion human beings in 1800 and 6.5 billion in 2002. During the same period, the number of people involved in farming dropped as the process became more automated. In the 1930s, 24 percent of the American population worked in agriculture compared to 1.5 percent in 2002; in 1940, each farm worker supplied 11 consumers, whereas in 2002, each worker supplied 90 consumers. The number of farms has also decreased, and their ownership is more concentrated. In the U.S., four companies kill 81 percent of cows, 73 percent of sheep, 57 percent of pigs, and produce 50 percent of chickens, cited as an example of "vertical integration" by the president of the U.S. National Farmers' Union. In 1967, there were one million pig farms in America; as of 2002, there were 114,000, with 80 million pigs (out of 95 million) killed each year on factory farms, according to the U.S. National Pork Producers Council. According to the Worldwatch Institute, 74 percent of the world's poultry, 43 percent of beef, and 68 percent of eggs are produced this way.

According to Denis Avery of the agribusiness funded Hudson Institute, Asia increased its consumption of pork by 18 million tons in the 1990s. As of 1997, the world had a stock of 900 million pigs, which Avery predicts will rise to 2.5 billion pigs by 2050. He told the College of Natural Resources at the University of California, Berkeley that three billion pigs will thereafter be needed annually to meet demand. He writes: "For the sake of the environment, we had better hope those hogs are raised in big, efficient confinement systems."

British Agricultural Revolution

The British agricultural revolution describes a period of agricultural development in Britain between the 16th century and the mid-19th century, which saw a massive increase in agricultural productivity and net output. This in turn supported unprecedented population growth, freeing up a significant percentage of the workforce, and thereby helped drive the Industrial Revolution. How this came about is not entirely clear. In recent decades, historians cited four key changes in agricultural practices, enclosure, mechanisation, four-field crop rotation, and selective breeding, and gave credit to a relatively few individuals.

Challenges and Issues

The challenges and issues of industrial agriculture for global and local society, for the industrial agriculture sector, for the individual industrial agriculture farm, and for animal rights include the costs and benefits of both current practices and proposed changes to those practices. This is a continuation of thousands of years of the invention and use of technologies in feeding ever growing populations.

When hunter-gatherers with growing populations depleted the stocks of game and wild foods across the Near East, they were forced to introduce agriculture. But agriculture brought much longer hours of work and a less rich diet than hunter-gatherers enjoyed. Further population growth among shifting slash-and-burn farmers led to shorter fallow periods, falling yields and soil erosion. Plowing and fertilizers were introduced to deal with these problems- but once again involved longer hours of work and degradation of soil resources (Boserup, The Conditions of Agricultural Growth, Allen and Unwin, 1965, expanded and updated in Population and Technology, Blackwell, 1980.).

While the point of industrial agriculture is lower cost products to create greater productivity thus a higher standard of living as measured by available goods and services, industrial methods have side effects both good and bad. Further, industrial agriculture is not some single indivisible thing, but instead is composed of numerous separate elements, each of which can be modified, and in fact is modified in response to market conditions, government regulation, and scientific advances. So the question then becomes for each specific element that goes into an industrial agriculture method or technique or process: What bad side effects are bad enough that the financial gain and good side effects are outweighed? Different interest groups not only reach different conclusions on this, but also recommend differing solutions, which then become factors in changing both market conditions and government regulations.

Society

The major challenges and issues faced by society concerning industrial agriculture include:

Maximising the benefits:

• Cheap and plentiful food

- Convenience for the consumer
- The contribution to our economy on many levels, from growers to harvesters to processors to sellers

while minimising the downsides:

o Environmental and social costs

o Damage to fisheries

o Cleanup of surface and groundwater polluted with animal waste

o Increased health risks from pesticides

o Increased ozone pollution via methane byproducts of animals

o Global warming from heavy use of fossil fuels.

Benefits

Population Growth

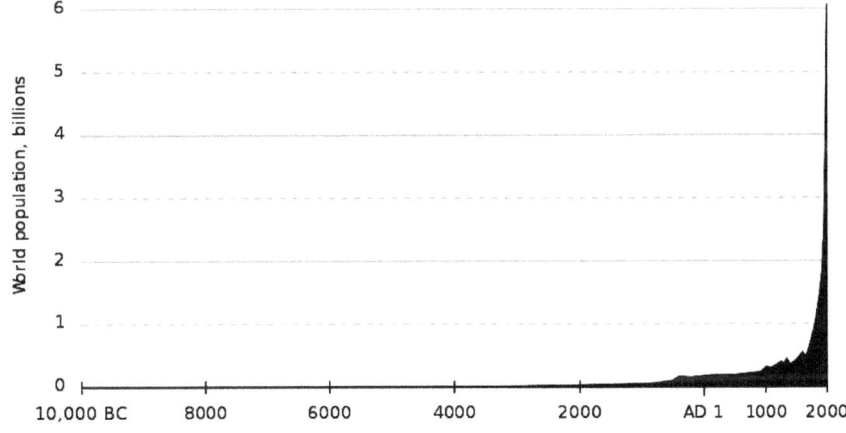

Population (est.) **10,000 BCE - 2000 CE.**

Very roughly:

- 30,000 years ago hunter-gatherer behaviour fed 6 million people
- 3,000 years ago primitive agriculture fed 60 million people
- 300 years ago intensive agriculture fed 600 million people
- Today industrial agriculture attempts to feed 6 billion people.

An example of industrial agriculture providing cheap and plentiful food is the U.S.'s "most successful program of agricultural development of any country in the world". Between 1930 and 2000 U.S. agricultural productivity (output divided by all inputs) rose by an average of about 2 percent annually causing food prices paid by consumers to decrease.

"The percentage of U.S. disposable income spent on food prepared at home decreased, from 22 percent as late as 1950 to 7 percent by the end of the century."

Liabilities

Environment

Industrial agriculture uses huge amounts of water, energy, and industrial chemicals; increasing pollution in the arable land, usable water and atmosphere. Herbicides, insecticides, fertilizers, and animal waste products are accumulating in ground and surface waters. "Many of the negative effects of industrial agriculture are remote from fields and farms. Nitrogen compounds from the Midwest, for example, travel down the Mississippi to degrade coastal fisheries in the Gulf of Mexico. But other adverse effects are showing up within agricultural production systems — for example, the rapidly developing resistance among pests is rendering our arsenal of herbicides and insecticides increasingly ineffective."

Social

A study done for the US. Office of Technology Assessment conducted by the UC Davis Macrosocial Accounting Project concluded that industrial agriculture is associated with substantial deterioration of human living conditions in nearby rural communities.

Animals

"Confined animal feeding operations" or "intensive livestock operations", can hold large numbers (some up to hundreds of thousands) of animals, often indoors. These animals are typically cows, hogs, turkeys, or chickens. The distinctive characteristics of such farms is the concentration of livestock in a given space. The aim of the operation is to produce as much meat, eggs, or milk at the lowest possible cost and with the greatest level of food safety.

Food and water is supplied in place, and artificial methods are often employed to maintain animal health and improve production, such as therapeutic use of antimicrobial agents, vitamin supplements and growth hormones. Growth hormones are not used in chicken meat production nor are they used in the European Union for any animal. In meat production, methods are also sometimes employed to control undesirable behaviours often related to stresses of being confined in restricted areas with other animals. More docile breeds are sought (with natural dominant behaviours bred out for example), physical restraints to stop interaction, such as individual cages for chickens, or animals physically modified, such as the de-beaking of chickens to reduce the harm of fighting. Weight gain is encouraged by the provision of plentiful supplies of food to animals breed for weight gain.

The designation "confined animal feeding operation" in the U.S. resulted from that country's 1972 Federal Clean Water Act, which was enacted to protect and restore lakes and rivers to a "fishable, swimmable" quality. The United States Environmental Protection Agency (EPA) identified certain animal feeding operations, along with many other types of industry, as point source polluters of groundwater. These operations were designated as CAFOs and subject to special anti-pollution regulation.

In 24 states in the U.S., isolated cases of groundwater contamination has been linked to CAFOs. For example, the ten million hogs in North Carolina generate 19 million tons of waste per year. The U.S. federal government acknowledges the waste disposal issue and requires that animal waste be stored in lagoons. These lagoons can be as large as 7.5 acres (30,000 m²). Lagoons not protected with an impermeable liner can leak waste into groundwater under some conditions, as can runoff from manure spread back onto fields as fertilizer in the case of an unforeseen heavy rainfall. A lagoon that burst in 1995 released 25 million gallons of nitrous sludge in North Carolina's New River. The spill allegedly killed eight to ten million fish.

The large concentration of animals, animal waste, and dead animals in a small space poses ethical issues to some consumers. Animal rights and animal welfare activists have charged that intensive animal rearing is cruel to animals. As they become more common, so do concerns about air pollution and ground water contamination, and the effects on human health of the pollution and the use of antibiotics and growth hormones.

According to the U.S. Centres for Disease Control and Prevention (CDC), farms on which animals are intensively reared can cause adverse health reactions in farm workers. Workers may develop acute and chronic lung disease, musculoskeletal injuries, and may catch infections that transmit from animals to human beings. These type of transmissions, however, and extremely rare, as zoonotic diseases are uncommon.

Crops

The projects within the Green Revolution spread technologies that had already existed, but had not been widely used outside of industrialised nations. These technologies included pesticides, irrigation projects, and synthetic nitrogen fertilizer.

The novel technological development of the Green Revolution was the production of what some referred to as "miracle seeds." Scientists created strains of maize, wheat, and rice that are generally referred to as HYVs or "high-yielding varieties." HYVs have an increased nitrogen-absorbing potential compared to other varieties. Since cereals that absorbed extra nitrogen would typically lodge, or fall over before harvest, semi-dwarfing genes were bred into their genomes. Norin 10 wheat, a variety developed by Orville Vogel from Japanese dwarf wheat varieties, was instrumental in developing Green Revolution wheat cultivars. IR8, the first widely implemented HYV rice to be developed by the International Rice Research Institute, was created through a cross between an Indonesian variety named "Peta" and a Chinese variety named "Dee Geo Woo Gen." With the availability of molecular genetics in Arabidopsis and rice the mutant genes responsible (*reduced height(rht), gibberellin insensitive (gai1)* and *slender rice (slr1)*) have been cloned and identified as cellular signalling components of gibberellic acid, a phytohormone involved in regulating stem growth via its effect on cell division. Stem growth in the mutant background is significantly reduced leading to the dwarf phenotype. Photosynthetic investment in the stem is reduced dramatically as the shorter plants are inherently more stable mechanically. Assimilates become redirected to grain production, amplifying in particular the effect of chemical fertilisers on commercial yield. HYVs significantly outperform traditional varieties in the presence of adequate irrigation, pesticides, and fertilizers. In the absence of these inputs, traditional varieties may outperform HYVs. One criticism of HYVs is that they were developed as F1 hybrids, meaning they need to be purchased by a farmer every season rather than saved from previous seasons, thus increasing a farmer's cost of production.

Sustainable Agriculture

The idea and practice of sustainable agriculture has arisen in response to the problems of industrial agriculture. Sustainable agriculture integrates three main goals: environmental stewardship, farm profitability, and prosperous farming communities. These goals have been defined by a variety of disciplines and may be looked at from the vantage point of the farmer or the consumer.

Organic Farming Methods

Organic farming methods combine some aspects of scientific knowledge and highly limited modern technology with traditional farming practices; accepting some of the methods of industrial agriculture while rejecting others. Organic methods rely on naturally occurring biological processes, which often take place over extended periods of time, and a holistic approach; while chemical-based farming focuses on immediate, isolated effects and reductionist strategies.

Integrated Multi-Trophic Aquaculture is an example of this holistic approach. Integrated Multi-Trophic Aquaculture (IMTA) is a practice in which the by-products (wastes) from one species are recycled to become inputs (fertilizers, food) for another. Fed aquaculture (e.g. fish, shrimp) is combined with inorganic extractive (e.g. seaweed) and organic extractive (e.g. shellfish) aquaculture to create balanced systems for environmental sustainability (biomitigation), economic stability (product diversification and risk reduction) and social acceptability (better management practices).

Dynamic Equilibrium

A dynamic equilibrium exists once a reversible reaction ceases to change its ratio of reactants/products, but substances move between the chemicals at an equal rate, meaning there is no net change. It is a particular example of a system in a steady state. In thermodynamics a closed system is in thermodynamic equilibrium when reactions occur at such rates that the composition of the mixture does not change with time. Reactions do in fact occur, sometimes vigorously, but to such an extent that changes in composition cannot be observed. Equilibrium constants can be expressed in terms of the rate constants for elementary reactions.

Examples

In a new bottle of cola the concentration of carbon dioxide in the liquid phase has a particular value. If half of the liquid is poured out and the bottle is sealed, carbon dioxide will leave the liquid phase at an ever decreasing rate and the partial pressure of carbon dioxide in the gas phase will increase until equilibrium is reached. At that point a molecule of CO_2 may leave the liquid phase, but then another molecule of CO_2 will pass from the gas to the liquid. At equilibrium the rate of loss of CO_2 is equal to the rate of gain. In this case, the equilibrium concentration of CO_2 in the liquid is given by Henry's law, which states that the solubility of a gas in a liquid is directly proportional to the partial pressure of that gas above the liquid. This relationship is written as;

$$p = kc$$

where k is a temperature-dependent constant, p is the partial pressure and c is the concentration of the dissolved gas in the liquid. Thus, the partial pressure of CO_2 in the gas has increased until Henry's law is obeyed. The concentration of carbon dioxide in the liquid has decreased and the drink has lost some of its fizz.

Henry's law may be derived by setting the chemical potentials of carbon dioxide in the two phases to be equal to each other. Equality of chemical potential defines chemical equilibrium. Other constants for dynamic equilibrium involving phase changes include partition coefficient and solubility product. Raoult's law defines the equilibrium vapour pressure of an ideal solution.

Dynamic equilibria can also exist in a homogeneous system. A simple example occurs with acid-base equilibria such as the "dissociation" of acetic acid, in aqueous solution.

$$CH_3CO_2H \square \quad CH_3CO_2^- + H^+$$

At equilibrium the concentration quotient, K, the acid dissociation constant, is constant (subject to some conditions)

$$K_c = \frac{[CH_3CO_2^-][H^+]}{[CH_3CO_2H]}$$

In this case, the forward reaction involves the liberation of some protons from acetic acid molecules and the backward reaction involves the formation of acetic acid molecules when an acetate ion accepts a proton. Equilibrium is attained when the sum of chemical potentials of the species on the left-hand side of the equilibrium expression is equal to the sum of chemical potentials of the species on the right-hand side. At the same time the rates of forward and backward reactions are equal to each other. Equilibria involving the formation of chemical complexes are also dynamic equilibria and concentrations are governed by the stability constants of complexes.

Dynamic equilibria can also occur in the gas phase as, for example, when nitrogen dioxide dimerizes.

$$2NO_2 \square \quad N_2O_4; K_P = \frac{P(N_2O_4)}{P(NO_2)^2}$$

In the gas phase, square brackets are not used as these indicate a concentration, instead a capitalised P is used to indicate partial pressure.

Relationship between equilibrium and rate constants

In a simple reaction such as the isomerization:

$$A \square \quad B$$

there are two reactions to consider, the forward reaction in which the species A is converted into B and the backward reaction in which B is converted into A. If both reactions are elementary reactions, then the rate of reaction is given by

$$\frac{d[A]}{dt} = -k_f [A]_t + k_b [B]_t$$

where k_f is the rate constant for the forward reaction and k_b is the rate constant for the backward reaction and the square brackets, [..] denote concentration. If only A is present at the beginning, time $t=0$, with a concentration $[A]_0$, the sum of the two concentrations, $[A]_t$ and $[B]_t$, at time t, will be equal to $[A]_0$.

$$\frac{d[A]}{dt} = -k_f [A]_t + k_b ([A]_0 - [A]_t)$$

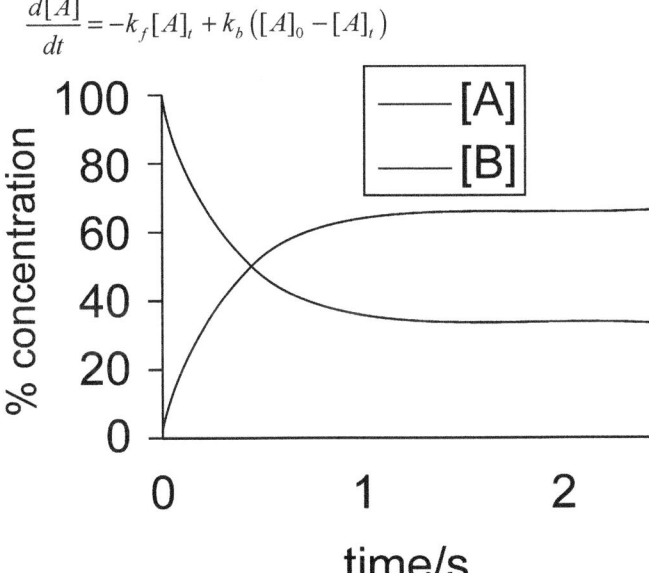

% concentrations of species in isomerization reaction. $k_f = 2$ s^{-1}, $k_r = 1$ s^{-1}

The solution to this differential equation is

$$[A]_t = \frac{k_b + k_f e^{-(k_f + k_b)t}}{k_f + k_b} [A]_0$$

and is illustrated at the right. As time tends towards infinity, the concentrations $[A]_t$ and $[B]_t$ tend towards constant values. Let t approach infinity, that is, $t'!\bullet$, in the expression above:

$$[A]_\infty = \frac{k_b}{k_f + k_b} [A]_0 ; [B]_\infty = \frac{k_f}{k_f + k_b} [A]_0$$

In practice, concentration changes will not be measurable after $t < \dfrac{10}{k_f + k_b}$. Since the concentrations do not change thereafter, they are, by definition, equilibrium concentrations. Now, the equilibrium constant for the reaction is defined as

$$K = \frac{[B]_{eq}}{[A]_{eq}}$$

It follows that the equilibrium constant is numerically equal to the quotient of the rate constants.

$$K = \frac{\dfrac{k_f}{k_f + k_b}[A]_0}{\dfrac{k_b}{k_f + k_b}[A]_0} = \frac{k_f}{k_b}$$

In general they may be more than one forward reaction and more than one backward reaction. Atkins states that, for a general reaction, the overall equilibrium constant is related to the rate constants of the elementary reactions by

$$K = \left(\frac{k_f}{k_b}\right)_1 \times \left(\frac{k_f}{k_b}\right)_2 \cdots$$

Environmental Economics

Central to environmental economics is the concept of market failure. Market failure means that markets fail to allocate resources efficiently. As stated by Hanley, Shogren, and White (2007) in their textbook *Environmental Economics*: "A market failure occurs when the market does not allocate scarce resources to generate the greatest social welfare. A wedge exists between what a private person does given market prices and what society might want him or her to do to protect the environment. Such a wedge implies wastefulness or economic inefficiency; resources can be reallocated to make at least one person better off without making anyone else worse off." Common forms of market failure include externalities, non-excludability and non-rivalry.

Externality: the basic idea is that an externality exists when a person makes a choice that affects other people that are not accounted for in the market price. For instance, a firm emitting pollution will typically not take into account the costs that its pollution imposes on others. As a result, pollution in excess of the 'socially efficient' level may occur. A classic definition influenced by Kenneth Arrow and James Meade is provided by Heller and Starrett (1976), who define an externality as "a situation in which the private economy lacks sufficient incentives to create a potential market in some good and the nonexistence of this market results in losses of Pareto efficiency." In economic terminology, externalities are examples of market failures, in which the unfettered market does not lead to an efficient outcome.

Common property and non-exclusion: When it is too costly to exclude people from access to an environmental resource for which there is rivalry, market allocation is likely to be inefficient. The challenges related with common property and non-exclusion have long been recognised.

Hardin's (1968) concept of the tragedy of the commons popularised the challenges involved in non-exclusion and common property. "Commons" refers to the environmental asset itself, "common property resource" or "common pool resource" refers to a property right regime that allows for some collective body to devise schemes to exclude others, thereby allowing the capture of future benefit streams; and "open-access" implies no ownership in the sense that property everyone owns nobody owns.

The basic problem is that if people ignore the scarcity value of the commons, they can end up expending too much effort, over harvesting a resource (e.g., a fishery). Hardin theorises that in the absence of restrictions, users of an open-access resource will use it more than if they had to pay for it and had exclusive rights, leading to environmental degradation. However, Ostrom's (1990) work on how people using real common property resources have worked to establish self-governing rules to reduce the risk of the tragedy of the commons.

Public goods and non-rivalry: Public goods are another type of market failure, in which the market price does not capture the social benefits of its provision. For example, protection from the risks of climate change is a public good since its provision is both non-rival and non-excludable. Non-rival means climate protection provided to one country does not reduce the level of protection to another country; non-excludable means it is too costly to exclude any one from receiving climate protection. A country's incentive to invest in carbon abatement is reduced because it can "free ride" off the efforts of other countries. Over a century ago, Swedish economist Knut Wicksell (1896) first discussed how public goods can be under-provided by the market because people might conceal their preferences for the good, but still enjoy the benefits without paying for them.

Valuation

Assessing the economic value of the environment is a major topic within the field. Use and indirect use are tangible benefits accruing from natural resources or ecosystem services. Non-use values include existence, option, and bequest values. For example, some people may value the existence of a diverse set of species, regardless of the effect of the loss of a species on ecosystem services. The existence of these species may have an option value, as there may be possibility of using it for some human purpose (certain plants may be researched for drugs). Individuals may value the ability to leave a pristine environment to their children.

Use and indirect use values can often be inferred from revealed behaviour, such as the cost of taking recreational trips or using hedonic methods in which values are estimated based on observed prices. Non-use values are usually estimated using stated preference methods such as contingent valuation or choice modelling. Contingent valuation typically takes the form of surveys in which people are asked how much they would pay to observe and recreate in the environment (willingness to pay) or their willingness to accept (WTA) compensation for the destruction of the environmental good. Hedonic pricing examines the effect the environment has on economic decisions through housing prices, travelling expenses, and payments to visit parks.

Solutions

Solutions advocated to correct such externalities include:

- *Environmental regulations.* Under this plan, the economic impact has to be estimated by the regulator. Usually this is done using cost-benefit analysis. There is a growing realisation that regulations (also known as "command and control" instruments) are not so distinct from economic instruments as is commonly asserted by proponents of environmental economics. E.g.1 regulations are enforced by fines, which operate as a form of tax if pollution rises above the threshold prescribed. E.g.2 pollution must be monitored and laws enforced, whether under a pollution tax regime or a regulatory regime. The main difference an environmental economist would argue exists between the two methods, however, is the total cost of the regulation. "Command and control" regulation often applies uniform emissions limits on polluters, even though each firm has different costs for emissions reductions. Some firms, in this system, can abate inexpensively, while others can only abate at high cost. Because of this, the total abatement has some expensive and some inexpensive efforts to abate. Environmental economic regulations find the cheapest emission abatement efforts first, then the more expensive methods second. E.g. as said earlier, trading, in the quota system, means a firm only abates if doing so would cost less than paying someone else to make the same reduction. This leads to a lower cost for the total abatement effort as a whole.

- *Quotas on pollution.* Often it is advocated that pollution reductions should be achieved by way of tradeable emissions permits, which if freely traded may ensure that reductions in pollution are achieved at least cost. In theory, if such tradeable quotas are allowed, then a firm would reduce its own pollution load only if doing so would cost less than paying someone else to make the same reduction. In practice, tradeable permits approaches have had some success, such as the U.S.'s sulphur dioxide trading program or the EU Emissions Trading Scheme, and interest in its application is spreading to other environmental problems.

- *Taxes and tariffs on pollution/Removal of "dirty subsidies."* Increasing the costs of polluting will discourage polluting, and will provide a "dynamic incentive," that is, the disincentive continues to operate even as pollution levels fall. A pollution tax that reduces pollution to the socially "optimal" level would be set at such a level that pollution occurs only if the benefits to society (for example, in form of greater production) exceeds the costs. Some advocate a major shift from taxation from income and sales taxes to tax on pollution- the so-called "green tax shift."

- *Better defined property rights.* The Coase Theorem states that assigning property rights will lead to an optimal solution, regardless of who receives them, if transaction costs are trivial and the number of parties negotiating is limited. For example, if people living near a factory had a right to clean air and water, or the factory had the right to pollute, then either the factory could pay those affected by the pollution or the people could pay the factory not to pollute. Or, citizens could take action themselves as they would if other property rights were violated. The US River Keepers Law of the 1880s was an early example, giving citizens downstream the right to end pollution upstream themselves if government itself did not act (an early example of bioregional democracy). Many markets for "pollution rights" have been created in the late twentieth century. The assertion that defining property rights is a solution is controversial within the field of environmental economics and environmental law and policy more broadly; in Anglo-American and many other legal systems, one has the right to carry out any action unless the law expressly proscribes it. Thus, property rights are already assigned (the factory that is polluting has a right to pollute).

Relationship to Other Fields

Environmental economics is related to ecological economics but there are differences. Most environmental economists have been trained as economists.

They apply the tools of economics to address environmental problems, many of which are related to so-called market failures—circumstances wherein the "invisible hand" of economics is unreliable. Most ecological economists have been trained as ecologists, but have expanded the scope of their work to consider the impacts of humans and their economic activity on ecological systems and services, and vice-versa.

This field takes as its premise that economics is a strict subfield of ecology. Ecological economics is sometimes described as taking a more pluralistic approach to environmental problems and focuses more explicitly on long-term environmental sustainability and issues of scale.

Environmental economics is viewed as more pragmatic in a price system; ecological economics as more idealistic in its attempts not use money as a primary arbiter of decisions. These two groups of specialists sometimes have conflicting views which may be traced to the different philosophical underpinnings.

Another context in which externalities apply is when globalisation permits one player in a market who is unconcerned with biodiversity to undercut prices of another who is- creating a "race to the bottom" in regulations and conservation.

This in turn may cause loss of natural capital with consequent erosion, water purity problems, diseases, desertification, and other outcomes which are not efficient in an economic sense. This concern is related to the subfield of sustainable development and its political relation, the anti-globalisation movement.

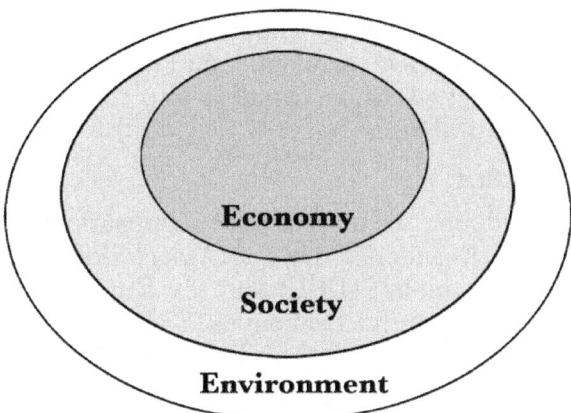

Three circles enclosed within one another showing how both economy and society are subsets of our planetary ecological system. This view is useful for correcting the misconception, sometimes drawn from the previous "three pillars" diagram that portions of social and economic systems can exist independently from the environment.

Environmental economics was once distinct from resource economics. Natural resource economics as a subfield began when the main concern of researchers was the optimal commercial exploitation of natural resource stocks. But resource managers and policy-makers eventually began to pay attention to the broader importance of natural resources (e.g. values of fish and trees beyond just their commercial exploitation;, externalities associated with mining). It is now difficult to distinguish "environmental" and "natural resource" economics as separate fields as the two became associated with sustainability. Many of the more radical green economists split off to work on an alternate political economy.

Environmental economics was a major influence for the theories of natural capitalism and environmental finance, which could be said to be two sub-branches of environmental economics concerned with resource conservation in production, and the value of biodiversity to humans, respectively. The theory of natural capitalism (Hawken, Lovins, Lovins) goes further than traditional environmental economics by envisioning a world where natural services are considered on par with physical capital.

The more radical Green economists reject neoclassical economics in favour of a new political economy beyond capitalism or communism that gives a greater emphasis to the interaction of the human economy and the natural environment, acknowledging that "economy is three-fifths of ecology"- Mike Nickerson. These more radical approaches would imply changes to money supply and likely also a bioregional democracy so that political, economic, and ecological "environmental limits" were all aligned, and not subject to the arbitrage normally possible under capitalism.

Professional Bodies

The main academic and professional organisations for the discipline of Environmental Economics are the Association of Environmental and Resource Economists (AERE) and the European Association for Environmental and Resource Economics (EAERE). The main academic and professional organisation for the discipline of Ecological Economics is the International Society for Ecological Economics (ISEE) and The Green Economics Institute for Green Economics is its international Professional body.

Chapter 8

Environmental Impact Assessment

An environmental impact assessment is an assessment of the possible positive or negative impact that a proposed project may have on the environment, together consisting of the natural, social and economic aspects.

The purpose of the assessment is to ensure that decision makers consider the ensuing environmental impacts when deciding whether to proceed with a project. The International Association for Impact Assessment (IAIA) defines an environmental impact assessment as "the process of identifying, predicting, evaluating and mitigating the biophysical, social, and other relevant effects of development proposals prior to major decisions being taken and commitments made." EIAs are unique in that they do not require adherence to a predetermined environmental outcome, but rather they require decision-makers to account for environmental values in their decisions and to justify those decisions in light of detailed environmental studies and public comments on the potential environmental impacts of the proposal.

EIAs began to be used in the 1960s as part of a rational decision making process. It involved a technical evaluation that would lead to objective decision making. EIA was made legislation in the US in the National Environmental Policy Act (NEPA) 1969. It has since evolved as it has been used increasingly in many countries around the world. As per Jay *et al.*(2006), EIA as it is practiced today, is being used as a decision aiding tool rather than decision making tool. There is growing dissent on the use of EIA as its influence on development decisions is limited and there is a view it is falling short of its full potential. There is a need for stronger foundation of EIA practice through training for practitioners, guidance on EIA practice and continuing research.

EIAs have often been criticized for having too narrow spatial and temporal scope. At present no procedure has been specified for determining a system boundary for the assessment. The system boundary refers to 'the spatial and temporal boundary of the proposal's effects'. This boundary is determined by the applicant and the lead assessor, but in practice, almost all EIAs address the direct, on-site effects alone.

However, as well as direct effects, developments cause a multitude of indirect effects through consumption of goods and services, production of building materials and machinery, additional land use for activities of various manufacturing and industrial services, mining of resources etc. The indirect effects of developments are often an order of magnitude higher than the direct effects assessed by EIA. Large proposals such as airports or ship yards cause wide ranging national as well as international environmental effects, which should be taken into consideration during the decision-making process.

Broadening the scope of EIA can also benefit threatened species conservation. Instead of concentrating on the direct effects of a proposed project on its local environment some EIAs used a landscape approach which focused on much broader relationships between the entire population of a species in question. As a result, an alternative that would cause least amount of negative effects to the population of that species as a whole, rather than the local subpopulation, can be identified and recommended by EIA.

There are various methods available to carry out EIAs, some are industry specific and some general methods:

- Industrial products- Product environmental life cycle analysis (LCA) is used for identifying and measuring the impact on the environment of industrial products. These EIAs consider technological activities used for various stages of the product: extraction of raw material for the product and for ancillary materials and equipment, through the production and use of the product, right up to the disposal of the product, the ancillary equipment and material.

- Genetically modified plants- There are specific methods available to perform EIAs of genetically modified plants. Some of the methods are GMP-RAM, INOVA etc.

- Fuzzy Arithmetic- EIA methods need specific parameters and variables to be measured to estimate values of impact indicators. However many of the environment impact properties cannot be measured on a scale e.g. landscape quality, lifestyle quality, social acceptance etc. and moreover these indicators are very subjective. Thus to assess the impacts we may need to take the help of information from similar EIAs, expert criteria, sensitivity of affected population etc. To treat this information, which is generally inaccurate, systematically, fuzzy arithmetic and approximate reasoning methods can be utilised. This is called as a fuzzy logic approach.

At the end of the project, an EIA should be followed by an audit. An EIA audit evaluates the performance of an EIA by comparing actual impacts to those that were predicted. The main objective of these audits is to make future EIAs more valid and effective. The two main considerations are:

- scientific- to check the accuracy of predictions and explain errors.

- management- to assess the success of mitigation in reducing impacts.

Some people believe that audits be performed as a rigorous scientific testing of the null hypotheses. While some believe in a simpler approach where you compare what actually occurred against the predictions in the EIA document.

After an EIA, the precautionary and polluter pays principles may be applied to prevent, limit, or require strict liability or insurance coverage to a project, based on its likely harms. Environmental impact assessments are sometimes controversial.

Application

Australia

The history of EIA in Australia could be linked to the enactment of the U.S. National Environment Policy Act (NEPA) in 1970, which made the preparation of environmental impact statements a requirement. In Australia, one might say that the EIA procedures were introduced at a State Level prior to that of the Commonwealth (Federal), with a majority of the states having divergent views to the Commonwealth. One of the pioneering states was New South Wales, whose State Pollution Control Commission issued EIA guidelines in 1974. At a Commonwealth (Federal) level, this was followed by passing of the Environment Protection (Impact of Proposals) Act in 1974. The Environment Protection and Biodiversity Conservation Act 1999 (EPBC) superseded the Environment Protection (Impact of Proposals) Act 1974 and is the current central piece for EIA in Australia on a Commonwealth (Federal) level. An important point to note is that this Commonwealth Act does not affect the validity of the States and Territories environmental and development assessments and approvals; rather the EPBC runs as a parallel to the State/Territory

Systems. Overlap between federal and state requirements is addressed via bilateral agreements or one off accreditation of state processes, as provided for in the EPBC Act.

The Commonwealth Level

The EPBC Act provides a legal framework to protect and manage nationally and internationally important flora, fauna, ecological communities and heritage places-defined in the EPBC Act as matters of 'national environmental significance'. Following are the eight matters of 'national environmental significance' to which the EPBC ACT applies :

- World Heritage sites;
- National Heritage places;
- RAMSAR wetlands of international significance;
- Listed threatened species and ecological communities;
- Migratory species protected under international agreements;
- The Commonwealth marine environment;
- Nuclear actions (including uranium mining); and
- National Heritage.

In addition to this, the EPBC Act aims at providing a streamlined national assessment and approval process for activities. These activities could be by the Commonwealth, or its agents, anywhere in the world or activities on Commonwealth land; and activities that are listed as having a 'significant impact' on matters of 'national environment significance'.

The EPBC Act comes into play when a person (a 'proponent') wants an action (often called a 'proposal' or 'project') assessed for environmental impacts under the EPBC Act, he or she must refer the project to the Department of Environment, Water, Heritage and the Arts (Australia).

This 'referral' is then released to the public, as well as relevant state, territory and Commonwealth ministers, for comment on whether the project is likely to have a significant impact on matters of national environmental significance. The Department of Environment, Water, Heritage and the Arts assess the process and makes recommendation to the minister or the delegate for the feasibility. The final discretion on the decision remains of the minister, which is not solely based on matters of 'national environmental significance' but also the consideration of social and economic impact of the project. The Australian Government environment minister cannot intervene in a proposal if it has no significant impact on one of the eight matters of 'national environmental significance' despite the fact that there may be other undesirable environmental impacts. This is primarily due to the division of powers between the States and the Federal government and due to which the Australian Government environment minister cannot overturn a state decision.

There are strict civil and criminal penalties for the breach of EPBC Act. Depending on the kind of breach, civil penalty (maximum) may go up to $550,000 for an individual and $5.5 million for a body corporate, or for criminal penalty (maximum) of seven years imprisonment and/or penalty of $46,200.

The State and Territory Level

Australian Capital Territory (ACT)

EIA in ACT is administered with the help of Part 4 of the Land (Planning and Environment) Act 1991 (Land Act) and Territory Plan (plan for land-use).

New South Wales (NSW)

In New South Wales, the Environment Planning Assessment Act 1979 (EPA) establishes three pathways for EIA. The first is under Part 3A of the Act where Major Projects require an environmental assessment. The second is under Part 4 of the Act dealing with development control. If a project does not require approval under Part 3A or Part 4 it is then potentially captured by the third pathway, Part 5 dealing with environment impact assessment.

Northern Territory (NT)

The EIA process in Northern Territory is chiefly administered under the Environmental Assessment Act (EEA). Although EEA is the primary tool for EIA in Northern Territory, there are further provisions for proposals in the Inquiries Act 1985 (NT).

Queensland (QLD)

There are four main EIA processes in Queensland. Firstly, under the Integrated Planning Act 1997 (IPA) for development projects other than mining. Secondly, under the Environmental Protection Act 1994 (EP Act) for some mining and petroleum activities. Thirdly, under the State Development and Public Works Organisation Act 1971 (State Development Act) for 'significant projects'. Finally, Environment Protection and Biodiversity Conservation Act 1999 (Cth) for 'controlled actions'.

South Australia (SA)

The local governing tool for EIA in South Australia is the Development Act 1993. There are three levels of assessment possible under the Act in the form of an environment impact statement (EIS), a public environmental report (PER) or a Development Report (DR).

Tasmania (TAS)

In Tasmania, an integrated system of legislation is used to govern development and approval process, this system is a mixture of the Environmental Management and Pollution Control Act 1994 (EMPCA), Land Use Planning and Approvals Act 1993 (LUPAA), State Policies and Projects Act 1993 (SPPA), and Resource Management and Planning Appeals Tribunal Act 1993.

Victoria (VIC)

The EIA process in Victoria is intertwined with the Environment Effects Act 1978 and the Ministerial Guidelines for Assessment of Environmental Effects (made under the s. 10 of the EE Act).

Western Australia (WA)

The Environmental Protection Act 1986 (Part 4) provides the legislative framework for the EIA process in Western Australia. The EPA Act oversees the planning and development proposals and assesses their likely impacts on the environment.

Canada

The *Canadian Environmental Assessment Act* (CEAA) is the legal basis for the federal environmental assessment (EA) process. CEAA came into force in 1995.

Legislative amendments were introduced in 2001 and came into force on October 30, 2003. EA is defined as a planning tool to identify, understand, assess and mitigate, where possible, the environmental effects of a project. Under the CEAA, all federal government departments and agencies are required to undertake an EA for projects relating to a physical work and for any proposed physical activity listed in the Inclusion List Regulations where it exercises one or more of the following CEAA triggers:

- Proposes or undertakes a project
- Grants money or any other form of financial assistance to a project
- Grants an interest in the land to enable a project to be carried out

- Exercises a regulatory duty in relation to a project, such as issuing a permit or license that is included in the Law List Regulations.

If a federal government department or agency exercises one or more of the above-mentioned triggers, it becomes a Responsible Authority (RA) under CEAA. As an RA, the federal department or agency in question must ensure that an EA is carried out in accordance with the CEAA and must consider the EA findings before a decision is made that could allow the project to proceed.

China

The Environmental Impact Assessment Law (EIA Law) requires an environmental impact assessment to be completed prior to project construction. However, if a developer completely ignores this requirement and builds a project without submitting an environmental impact statement, the only penalty is that the environmental protection bureau (EPB) may require the developer to do a make-up environmental assessment.

If the developer does not complete this make-up assessment within the designated time, only then is the EPB authorised to fine the developer. Even so, the possible fine is capped at a maximum of about US$25,000, a fraction of the overall cost of most major projects. The lack of more stringent enforcement mechanisms has resulted in a significant percentage of projects not completing legally required environmental impact assessments prior to construction.

China's State Environmental Protection Administration (SEPA) used the legislation to halt 30 projects in 2004, including three hydro-power plants under the Three Gorges Project Company. Although one month later (Note as a point of reference, that the typical EIA for a major project in the USA takes one to two years.), most of the 30 halted projects resumed their construction, reportedly having passed the environmental assessment, the fact that these key projects' construction was ever suspended was notable.

A joint investigation by SEPA and the Ministry of Land and Resources in 2004 showed that 30-40% of the mining construction projects went through the procedure of environment impact assessment as required, while in some areas only 6-7% did so. This partly explains why China has witnessed so many mining accidents in recent years.

SEPA alone cannot guarantee the full enforcement of environmental laws and regulations, observed Professor Wang Canfa, director of the centre to help environmental victims at China University of Political Science and Law. In fact, according to Wang, the rate of China's environmental laws and regulations that are actually enforced is estimated to be barely 10%.

Egypt

Environmental Impact Assessment (EIA) EIA is implemented in Egypt under the umbrella of the Ministry of state for environmental affairs. The Egyptian Environmental Affairs Agency (EEAA) is responsible for the EIA services.

In June 1997, the responsibility of Egypt's first full time Minister of State for Environmental Affairs was assigned as stated in the Presidential Decree no.275/1997. From thereon, the new ministry has focused, in close collaboration with the national and international development partners, on defining environmental policies, setting priorities and implementing initiatives within a context of sustainable development.

According to the Law 4/1994 for the Protection of the Environment, the Egyptian Environmental Affairs Agency (EEAA) was restructured with the new mandate to substitute the institution initially established in 1982. At the central level, EEAA represents the executive arm of the Ministry.

The purpose of EIA is to ensure the protection and conservation of the environment and natural resources including human health aspects against uncontrolled development. The long-term objective is to ensure a sustainable economic development that meets present needs without compromising future generations ability to meet their own needs. EIA is an important tool in the integrated environmental management approach.

EIA must be performed for new establishments or projects and for expansions or renovations of existing establishments according to the Law for the Environment.

EU

The European Union has established a mix of mandatory and discretionary procedures to assess environmental impacts. European Union Directive (85/337/EEC) on Environmental Impact Assessments (known as the *EIA Directive*) was first introduced in 1985 and was amended in 1997. The directive was amended again in 2003, following EU signature of the 1998 Aarhus Convention. In 2001, the issue was enlarged to the assessment of plans and programmes by the so called *Strategic Environmental Assessment (SEA) Directive* (2001/42/EC), which is now in force. Under the EU directive, an EIA must provide certain information to comply. There are seven key areas that are required:

1. Description of the project
   - Description of actual project and site description
   - Break the project down into its key components, i.e. construction, operations, decommissioning
   - For each component list all of the sources of environmental disturbance
   - For each component all the inputs and outputs must be listed, e.g., air pollution, noise, hydrology

2. Alternatives that have been considered
   - Examine alternatives that have been considered
   - Example: in a biomass power station, will the fuel be sourced locally or nationally?

3. Description of the environment
   - List of all aspects of the environment that may be affected by the development
   - Example: populations, fauna, flora, air, soil, water, humans, landscape, cultural heritage
   - This section is best carried out with the help of local experts, e.g. the RSPB in the UK

4. Description of the significant effects on the environment
   - The word significant is crucial here as the definition can vary
   - 'Significant' needs to be defined
   - The most frequent method used here is use of the Leopold matrix
   - The matrix is a tool used in the systematic examination of potential interactions
   - Example: in a windfarm development a significant impact may be collisions with birds

5. Mitigation
   - This is where EIA is most useful
   - Once section 4 has been completed it will be obvious where the impacts will be greatest
   - Using this information ways to avoid negative impacts should be developed

- Best working with the developer with this section as they know the project best
- Using the windfarm example again construction could be out of bird nesting seasons

6. Non-technical summary (EIS)

- The EIA will be in the public domain and be used in the decision making process
- It is important that the information is available to the public
- This section is a summary that does not include jargon or complicated diagrams
- It should be understood by the informed lay-person

7. Lack of know-how/technical difficulties

- This section is to advise any areas of weakness in knowledge
- It can be used to focus areas of future research
- Some developers see the EIA as a starting block for poor environmental management.

The Netherlands

EIA was implemented in Dutch legislation on September 1, 1987. The categories of projects that require an EIA are summarised in Dutch legislation, the Wet milieubeheer. The use of thresholds for activities makes sure that EIA is obligatory for those activities that may have considerable impacts on the environment.

For projects and plans that fit these criteria, a EIA report is required. The EIA report defines a.o. the proposed initiative, it makes clear the impact of that initiative on the environment and compares this with the impact of possible alternatives with less a negative impact.

Hong Kong

EIA in Hong Kong, since 1998, is regulated by the *Environmental Impact Assessment Ordinance 1997*.

The original proposal to construct the Lok Ma Chau Spur Line overground across the Long Valley failed to get through EIA, and the KCR Corporation had to change its plan and build the railway underground. In April 2011, the EIA of the Hong Kong section of the Hong Kong-Zhuhai-Macau Bridge was found to have breached the ordinance, and was declared unlawful. The appeal by the government was allowed in September 2011. However, it was estimated that this EIA court case had increased the construction cost of the Hong Kong section of the bridge by HK$6.5 billion in money-of-the-day prices.

India

The Ministry of Environment and Forests (MoEF) of India have been in a great effort in Environmental Impact Assessment in India. The main laws in nation are Water Act(1974), The Indian Wildlife (Protection) Act (1972), The Air (Prevention and Control of Pollution) Act (1981) and The Environment (Protection) Act (1986). The responsible body for this is Central Pollution Control Board. Environmental Impact Assessment (EIA) studies need a significant amount of primary and secondary environmental data. The primary data are those which need to be collected in the field to define the status of environment (like air quality data, water quality data etc.). The secondary data are those data which have been collected over the years and can be used to understand the existing environmental scenario of the study area. The environmental impact assessment (EIA) studies are conducted over a short period of time and therefore the understanding the environmental trends based on few months of primary data has its own limitations. Ideally, the primary data has to be considered along with the secondary data for complete understanding of the existing environmental status of the area. In many EIA studies, the secondary data needs could be as high as 80% of the total data requirement. EIC is the repository of one stop secondary data source for environmental impact assessment in India.

The Environmental Impact Assessment (EIA) experience in India indicates that the lack of timely availability of reliable and authentic environmental data has been a major bottle neck in achieving the full benefits of EIA. The environment being a multi-disciplinary subject, a multitude of agencies is involved in collection of environmental data. However, there is no single organisation in India which tracks the data available amongst these agencies and makes it available in one place, in a form and manner required by practitioners in the field of environmental impact assessment in India. Further, the environmental data is not available in value added forms that can enhance the quality of the EIA. This in turn adversely affects the time and efforts required for conducting the environmental impact assessments (EIAs) by project proponents and also timely environmental clearances by the regulators. With this background, Environmental Information Centre (EIC) has been set up to serve as a professionally managed clearing house of environmental information that can be used by

MoEF, project proponents, consultants, NGOs and other stakeholders involved in the process of environmental impact assessment in India. EIC caters to the need of creating and disseminating of organised environmental data for various developmental initiatives all over the country.

EIC stores data in GIS format and makes it available to all environmental impact assessment studies and to EIA stakeholders in a cost effective and timely manner.

Malaysia

In Malaysia, Section 34A, Environmental Quality Act, 1974 requires developments that have significant impact to the environment are required to conduct the Environmental impact assessment.

Nepal

In Nepal, EIA has been integrated in major development projects since the early 1980s. In the planning history of Nepal, the sixth plan (1980–85), for the first time, recognised the need for EIA with the establishment of Environmental Impact Study Project (EISP) under the Department of Soil Conservation in 1982 to develop necessary instruments for integration of EIA in infrastructure development projects. However, the government of Nepal enunciated environment conservation related policies in the seventh plan (NPC, 1985–1990). In order to enforce this policy and make necessary arrangements, a series of guidelines were developed, thereby incorporating the elements of environmental factors right from the project formulation stage of the development plans and projects and to avoid or minimise adverse effects on the ecological system. In addition, it has also emphasized that EIAs of industry, tourism, water resources, transportation, urbanisation, agriculture, forest and other developmental projects be conducted.

In Nepal, the government's Environmental Impact Assessment Guideline of 1993 inspired the enactment of the Environment Protection Act (EPA) of 1997 and the Environment Protection Rules (EPR) of 1997(EPA and EPR have been enforced since 24 and 26 June 1997 respectively in Nepal) to internalizing the environmental assessment system. The process institutionalized the EIA process in development proposals and enactment, which makes the integration of IEE and EIA legally binding to the prescribed projects. The projects, requiring EIA or IEE, are included in Schedules 1 and 2 of the EPR, 1997 (GoN/MoLJPA 1997). Progresses were made in the Environmental protection issue during the 8th five year plan (1992–1997). The following development in Environmental protection were achieved during that time:

- Formulation of Environmental Protection Act 1997
- Establishment of Ministry of Environment
- Development of National Environmental Policies and Action Plan, EIA guidelines developed
- Consideration of environmental concerns in hydropower projects
- Development of industrial, irrigation and agricultural policies that undertook environmental concerns

Source: Bhatta R. and Khanal S. 2010. African Journal of Environmental Science and Technology Vol. 4(9).

New Zealand

In New Zealand, EIA is usually referred to as *Assessment of Environmental Effects* (AEE). The first use of EIA's dates back to a Cabinet minute passed in 1974 called Environmental Protection and Enhancement Procedures. This had no legal force and only related to the activities of government departments. When the Resource Management Act was passed in 1991, an EIA was required as part of a resource consent application. Section 88 of the Act specifies that the AEE must include "such detail as corresponds with the scale and significance of the effects that the activity may have on the environment". While there is no duty to consult any person when making a resource consent application (Sections 36A and Schedule 4), proof of consultation is almost certain to be required by local councils when making a decision about whether or not to publicly notifiy the consent application under Section 93.

Russian Federation

Russia holds the world's largest natural gas reserves, the second largest coal reserves, and the eighth largest oil reserves. Russia is also the world's largest exporter of natural gas, the second largest oil exporter and the third largest energy consumer. As of 2004, the state authority responsible for conducting the State EIA in Russia has been split between two Federal bodies: 1) Federal service for monitoring the use of natural resources – a part of the Russian Ministry for Natural Resources and Environment and 2) Federal Service for Ecological, Technological and Nuclear Control. The two main pieces of environmental legislation in Russia are: The Federal Law 'On Ecological Expertise, 1995 and the 'Regulations on Assessment of Impact from Intended Business and Other Activity on Environment in the Russian Federation, 2000.

Federal Service for Monitoring the use of Natural Resources

In 2006, the parliament committee on ecology in conjunction with the Ministry for Natural Resources and Environment, created a working group to prepare a number of amendments to existing legislation in order to cover such topics as stringent project documentation for building of potentially environmentally damaging objects as well as building of projects on the territory of protected areas. There has been some success in this area, as evidenced from abandonment of plans to construct a gas pipeline through the only remaining habitat of the critically endangered Amur leopard in the Russian Far East.

Federal Service for Ecological, Technological and Nuclear Control

The government's decision to hand over control over several important procedures, including state EIA in the field of all types of energy projects, to the Federal Service for Ecological, Technological and Nuclear Control had caused a major controversy and criticism from environmental groups that blamed the government for giving nuclear power industry control over the state EIA.

Not surprisingly the main problem concerning State EIA in Russia is the clear differentiation of jurisdiction between the two above-mentioned Federal bodies.

Sri Lanka

Environmental Impact Assessments

One popular approach to assist in smart growth in democratic countries is for lawmakers to require prospective developers to prepare environmental impact assessments of their plans as a condition for state and/or local governments to go for Environmental Impact Assessments

These reports often indicate how significant impacts generated by the development will be mitigated- the cost of which is usually paid by the developer. These assessments are frequently controversial. Conservationists, neighbourhood advocacy groups and NIMBYs are often skeptical about such impact reports, even when they are prepared by independent agencies and subsequently approved by the decision makers rather than the promoters. Conversely, developers will sometimes strongly resist being required to implement the mitigation measures required by the local government as they may be quite costly.

These assessments are frequently controversial. Conservationists, neighbourhood advocacy groups and NIMBYs are often skeptical about such impact reports, even when they are prepared by independent agencies and subsequently approved by the decision makers rather than the promoters. Conversely, developers will sometimes strongly resist being required to implement the mitigation measures required by the local government as they may be quite costly.

The importance of the Environmental Impact Assessment as an effective tool for the purpose of integrating environmental considerations with development planning is highly recognised in Sri Lanka. The application of this technique is considered as a means of ensuring that the likely effects of new development projects on the environment are fully understood and taken into account before development is allowed to proceed. The importance of this management tool to foresee potential environmental impacts and problems caused by proposed projects and its use as a mean to make project more suitable to the environment are highly appreciated.

United States

Under United States environmental law an Environmental Assessment (EA) is compiled to determine the need for an *Environmental Impact Statement* (EIS), and originated in the National Environmental Policy Act (NEPA), enacted in 1969. Certain actions of federal agencies must be preceded by an EA or EIS. Contrary to a widespread misconception, NEPA does not prohibit the federal government or its licensees/permittees from harming the environment, nor does it specify any penalty if the EA or EIS turns out to be inaccurate, intentionally or otherwise. NEPA requires that plausible statements as to the prospective impacts be disclosed in advance. The purpose of NEPA process is to ensure that the decision maker is fully informed of the environmental aspects and consequences prior to making the final decision.

An *Environmental Assessment (EA)* is an environmental analysis prepared pursuant to the National Environmental Policy Act to determine whether a federal action would significantly affect the environment and thus require a more detailed *Environmental Impact Statement (EIS)*. The release of an Environmental Assessment results in either a *Finding of No Significant Impact (FONSI)* or an *Environmental Impact Statement (EIS)*.

The Environmental Assessment is a concise public document prepared by the federal action agency that serves to:

1. briefly provide sufficient evidence and analysis for determining whether to prepare an EIS or a Finding of No Significant Impact (FONSI)

2. Demonstrate compliance with the act when no EIS is required

3. facilitate the preparation of a EIS when a FONSI cannot be demonstrated.

The Environmental Assessment includes a brief discussion of the purpose and need of the proposal and of its alternatives as required by CFR 102(2)(E), of the environmental impacts of the proposed actions and alternatives, as well as a listing of agencies and stakeholders consulted. The action agency must approve an EA before it is made available to the public. The EA is made public through notices of availability by local, state, or regional clearing houses, newspapers, etc. There is a 15-30 day review period required for an Environmental Assessment, while the document is made available for public commentary.

An agency will release either a *Draft Environmental Assessment* (Draft EA) or a *Draft Environmental Impact Statement* (DEIS) for comment. Interested parties and the general public have the opportunity to comment on the draft, after which the agency will address all comments received and prepare a decision document, either a FONSI, Notice of Intent (NOI) to prepare a EIS or a Record of Decision for an EIS. The agency will then approve the " Final Environmental Assessment" (Final EA) or *Final Environmental Impact Statement* (FEIS). Commenting on the Draft EA is typically done in writing or email, submitted to the lead agency as defined in the Notice of Availability. Draft EIS's require public hearings, so comments can be made in person, as well as in writing. Occasionally, the agency will later release a "Supplemental Environmental Assessment " (Supplemental EA) or a *Supplemental Environmental Impact Statement* (SEIS), if the project parameters or environmental conditions change substantially after the issuance of the FONSI or ROD.

The adequacy of an EIS can be challenged in federal court. Major proposed projects have been blocked because of an agency's failure to prepare an acceptable EIS. One prominent example was the Westway landfill and highway development in and along the Hudson River in New York City. Another prominent case involved the Sierra Club suing the Nevada Department of Transportation over its denial of Sierra Club's request to issue a supplemental EIS addressing air emissions of particulate matter and hazardous air pollutants in the case of widening US Highway 95 through Las Vegas. The case reached the United States Court of Appeals for the Ninth Circuit, which led to construction on the highway being halted until the court's final decision. The case was settled prior to the court's final decision.

Several state governments that have adopted "little NEPAs," state laws imposing EIS requirements for particular state actions. Some those state laws such as ()the California Environmental Quality Act refer to the required environmental impact studies as environmental impact reports.

The structure of a generic Environmental Assessment is as follows:

1. Summary
2. Introduction
   - Structure
   - Background
   - Purpose and Need for Action
   - Proposed Action
   - Decision Framework
   - Public Involvement
   - Issues
3. Alternatives, including the Proposed Action
   - Alternatives
   - Mitigation Common to All Alternatives
   - Comparison of Alternatives
   4. Environmental Consequences
5. Consultation and Coordination.

These variety of state requirements are yielding voluminous data not just upon impacts of individual projects, but also to elucidate scientific areas that had not been sufficiently researched. For example, in a seemingly routine *Environmental Impact Report* for the city of Monterey, California, information came to light that led to the official federal endangered species listing of Hickman's potentilla, a rare coastal wildflower.

Transboundary Application

Environmental threats do not respect national borders. International pollution can have detrimental effects on the atmosphere, oceans, rivers, aquifers, farmland, the weather and biodiversity. Global climate change is transnational. Specific pollution threats include acid rain, radioactive contamination, debris in outer space, stratospheric ozone depletion and toxic oil spills. The Chernobyl disaster, precipitated by a nuclear accident on April 26, 1986, is a stark reminder of the devastating effects of transboundary nuclear pollution.

Environmental protection is inherently a cross-border issue and has led to the creation of transnational regulation via multilateral and bilateral treaties. The United Nations Conference on the Human Environment (UNCHE or Stockholm Conference) held in Stockholm in 1972 and the United Nations Conference on the Environment and Development (UNCED or Rio Summit, Rio Conference, or Earth Summit) held in Rio de Janeiro in 1992 were key in the creation of about 1,000 international instruments that include at least some provisions related to the environment and its protection. The United Nations Economic Commission for Europe's Convention on Environmental Impact Assessment in a Transboundary Context was negotiated to provide an international legal framework for transboundary EIA. However, as there is no universal legislature or administration with a comprehensive mandate, most international treaties exist parallel to one another and are further developed without the benefit of consideration being given to potential conflicts with other agreements. There is also the issue of international enforcement. This has led to duplications and failures, in part due to an inability to enforce agreements. An example is the failure of many international fisheries regimes to restrict harvesting practises.

Annexed Projects

All projects are either classified as Annex 1 or Annex 2 projects. Those lying in Annex 1 are large scale developments such as motorways, chemical works, bridges, powerstations etc. These always require an EIA under the Environmental Impact Assessment Directive (85/337/EEC) Annex 2 projects are smaller in scale such as small installments, and maintenance works. These do not require an EIA.

Farmer Field School

The Farmer Field School (FFS) is a group-based learning process that has been used by a number of governments, NGOs and international agencies to promote Integrated Pest Management (IPM). The first FFS were designed and managed by the UN Food and Agriculture Organisation in Indonesia in 1989 since when more than two million farmers across Asia have participated in this type of learning. The Farmer Field School brings together concepts and methods from agroecology, experiential education and community development. As a result, hundreds of thousands of rice farmers in countries such as China, Indonesia, Philippines and Vietnam have been able to reduce the use of pesticides and improve the sustainability of crop yields. The FFS has produced other developmental benefits that are broadly described as 'empowerment': FFS alumni in a number of countries are involved in a wide-range of self-directed activities including research, training, marketing and advocacy.

Origins of the Farmer Field School

Almost one third of the world's population are members of farming households in Asia. Most of these farming families are small holders. Forty years ago, the Green Revolution was launched with the aim of improving the productivity of small farmers. By improving access to water, improved varieties, and other inputs, the Green Revolution helped to double average rice yields between the 1960s and the 1990s. The Green Revolution offered hybrid seeds with reduced reproduction capacity to farmers traditionally practicing seed saving. The farmers had to buy new seeds for every season since the hybrid crop's qualities wore off after the second or third generation of seed. In effect many farmers became dependent on repurchasing something that was once free. The new varieties required the optimal growing conditions, thus if irrigation was inaccessible (as it is in many areas), if you weren't able to take out loans for excess to fertilizer and pesticide and seed, then the plants would not grow as well as the local domesticated variates adapted to that environment. Farmers unable to pay off their loans in these areas committed suicide by poisoning themselves with pesticides. In just the Indian state of Chattisgarh over 1,500 farmers committed suicide. These farmer suicides have, however, mostly been linked to the introduction of the Genetically modified Bt-cotton in India. A report by the International Food Policy Research Institute (IFPRI) published a report in 2008 concludes, among other things, with writing that "Bt cotton is neither a necessary nor a sufficient condition for the occurrence of farmer suicides".

During the 1970s it became increasingly apparent that pest resistance and resurgence caused by the indiscriminate use of insecticides posed an immediate threat to the gains of the Green Revolution. At the same time, research was being conducted that demonstrated the viability of biological control of major rice pests. However, gaps still existed between the science generated in research institutions and common farmer practice conditioned by years of aggressive promotion of pesticide use. Over the ensuing years, a number of approaches were tried to bring integrated pest management (IPM) to small farmers- particularly rice farmers- in Asia, with mixed results. Some experts claimed that the principles of IPM were too complex for small farmers to master, and that centrally-designed messages were still the only way to convince farmers to change their practices. By the end of the 1980s, a new approach to farmer training emerged in Indonesia called the 'Farmer Field School' (FFS). The broad problem which these field schools were designed to address was a lack of knowledge among Asian farmers relating to agroecology, particularly the relationship between insect pests and beneficial insects.

The implementation of projects using the FFS approach led to a deeper understanding of the problem and its causes. It was recognised that sustainable agricultural development required more than just the acquisition of ecological knowledge by individual farmers. It also required the development of a capability for generating, adapting and extending this knowledge within farming communities. The weakness of this capability in most farming communities is itself an important problem; one which has often been exacerbated by earlier agricultural development programmes that fostered a dependency on external sources of expertise. This deeper understanding of the problem was first recognised by farmers in Indonesia who graduated from FFS but realised there was more they could do to improve rural livelihoods. They started to organise new groups, alliances, networks and associations, and became involved in planning and implementing their own interventions. These interventions were highly diverse, ranging from research and training, to marketing and advocacy work. In response to the activities of these groups, IPM projects started to support the idea of 'Community IPM', which gave considerable attention to organisational issues rather than focussing solely on technological and educational aspects of IPM.

Description of a Typical Farmer Field School

The Farmer Field School (FFS) is a group-based learning process. During the FFS, farmers carried out experiential learning activities that helped them understand the ecology of their rice fields. These activities involve simple experiments, regular field observations and group analysis. The knowledge gained from these activities enables participants to make their own locally-specific decisions about crop management practices. This approach represents a radical departure from earlier agricultural extension programmes, in which farmers were expected to adopt generalised recommendations that had been formulated by specialists from outside the community. The basic features of a typical rice IPM Farmer Field School are as follows:

- The IPM Field School is field based and lasts for a full cropping season.
- A rice FFS meets once a week with a total number of meetings that might range from at least 10 up to 16 meetings.

- The primary learning material at a Farmers Field School is the rice field.

- The Field School meeting place is close to the learning plots often in a farmer's home and sometimes beneath a convenient tree.

- FFS educational methods are experiential, participatory, and learner centred.

- Each FFS meeting includes at least three activities: the agro-ecosystem analysis, a "special topic", and a group dynamics activity.

- In every FFS, participants conduct a study comparing IPM with non-IPM treated plots.

- An FFS often includes several additional field studies depending on local field problems.

- Between 25 and 30 farmers participate in a FFS. Participants learn together in small groups of five to maximise participation.

- All FFSs include a Field Day in which farmers make presentations about IPM and the results of their studies.

- A pre- and post-test is conducted as part of every Field School for diagnostic purposes and for determining follow-up activities.

- The facilitators of FFS's undergo intensive season-long residential training to prepare them for organising and conducting Field Schools.

- Preparation meetings precede an FFS to determine needs, recruit participants, and develop a learning contract.

- Final meetings of the FFS often include planning for follow-up activities.

Although Farmer Field Schools were designed to promote IPM, empowerment has an essential feature from the beginning. The curriculum of the FFS was built on the assumption that farmers could only implement IPM once they had acquired the ability to carry out their own analysis, make their own decisions and organise their own activities. The empowerment process, rather than the adoption of specific IPM techniques, is what produces many of the developmental benefits of the FFS.

FAO Support for Farmer Field Schools in Asia

The first IPM Farmer Field Schools were designed and managed in 1989 by experts working for the UN Food and Agriculture Organisation (FAO) in Indonesia. This was not, however, the first attempt made by FAO to extend IPM techniques to farmers in South East Asia. The FAO *Intercountry Programme for the Development and Application of Integrated Pest Control in Rice in South and South-East Asia* started in 1980, building on the experience of the International Rice Research Institute (IRRI) and Bureau of Plant Industry in the Philippines. Over the following two decades the Intercountry Programme (ICP) played a leading role in the promotion of rice IPM in Asia, giving rise to numerous other projects and programmes. By the time of completion in 2002, the ICP had a cumulative budget of $45 million, which had been spent on training activities in 12 countries (Bangladesh, Cambodia, China, India, Indonesia, Laos, Malaysia, Nepal, Philippines, Sri Lanka, Thailand, Vietnam). The ICP was not the only IPM programme supported by FAO during this period. Essential to the development of the FFS was a National IPM Programme in Indonesia, which ran between 1989 and 2000, funded by the United States ($ 25 million grant), World Bank ($ 37 million loan) and the Government ($ 14 million). FAO provided technical assistance to the National IPM Programme through a team of experts based in Indonesia, with back-stopping from the ICP. National projects were also developed and supported by FAO on a smaller scale in Bangladesh, Cambodia, China and Nepal. Additionally, the ICP launched 'spin-off' regional programmes focusing on IPM in cotton and vegetables. In total, during the 15-year period between 1989 and 2004, approximately $100 million in grants were allocated to IPM projects in Asia that used the FFS approach under the guidance of FAO

Costs and Benefits of the Farmer Field School

There are two major reasons why it is difficult to make generalisations about the costs and benefits of IPM field schools. Firstly, there is a lack of agreement about what factors should be taken into account on both sides of the cost-benefit equation. Regarding benefits, should we limit ourselves to measuring yields and pesticide savings, or should we also take account of improvements in public health and the consequences of farmers becoming better organised? Regarding costs, should we limit ourselves to the expenses involved in running field schools, or should we also take account of the wider costs of training extension staff and managing IPM programmes.

Secondly, there is a high degree of variation in the value of individual factors. The cost of conducting a season-long field school for 25 farmers has ranged from $150 to $1,000 depending on the country and the organisation. In some cases, the graduates of FFS have saved $40 per hectare per season by eliminating pesticides without any loss of yield. In other cases, graduates did not experience any savings because they were not previously using any pesticides, but yields increased by as much as 25% as a result of adopting other practices learnt during the FFS, such as improved varieties, better water management and enhanced plant nutrition. The conceptual and methodological problems associated with assessing the impact of IPM field schools have resulted in disagreements among experts about the advantages of this intervention. One widely circulated paper written by World Bank economists has questioned the benefit of 'sending farmers back to school'. By contrast, a meta-analysis of 25 impact studies commissioned by FAO concluded:

> *The majority of studies... reported substantial and consistent reductions in pesticide use attributable to the effect of training. In a number of cases, there was also a convincing increase in yield due to training....*
>
> *A number of studies described broader, developmental impacts of training.... Results demonstrated remarkable, widespread and lasting developmental impacts. It was found that the FFS stimulated continued learning, and that it strengthened social and political skills, which apparently prompted a range of local activities, relationships and policies related to improved agro-ecosystem management.*

Due to differences in motivation, scope of analysis and methodology, it is unlikely that experts from the World Bank and FAO will reach agreement on the advantages and disadvantages of the IPM field school in the near future.

Despite the arguments amongst economists and policy makers, there has been widespread enthusiasm for IPM and FFS among farmers and development practitioners in a number of Asian countries. Participation in FFS has always been voluntary, and none of the IPM projects and programmes supported by FAO provided financial incentives to participants. On the contrary, participation in FFS has always involved a considerable cost in terms of time and effort. Despite these costs, two million farmers decided to participate. In most countries, the demand for places on a field school has been ahead of supply, and drop-out rates have been very low. Furthermore, there are many examples of farmers who decided to train other members of their community and continue working as a group after the training came to an end.

Bibliography

Asaithambi S. : *Economics of Ground Water Management in India*, Abhijeet, Delhi, 2008.

Bandyopadhyay, P C : *Breeding and Crop Production*, Gene Tech Books, Delhi, 2007.

Bennett, Hugh Hammond : *Elements of Soil Conservation*, Biotech Books, Delhi, 2009.

Bhakar S.R. : *Ground Water Hydrology : Theory and Practice*, Agrotech, Delhi, 2009.

Biswas Asit K. : *Integrated Water Resources Management in South and South-East Asia*, , Oxford University Press, Delhi, 2001.

Bora, K.K. : *Agros Dictionary of Plant Physiology and Biochemistry*, Agrobios, Delhi, 2001.

Boyer , J.S.: *Measuring the Water Status of Plants and Soils,* Academic Press, N.Y., 1995.

Chadha K. L. and Pareek O. P.: *Advances in Horticulture: Fruit Crops,* New Delhi, Malhotra Publishing House, 1993.

Chahal, S.S. : *Achievements and Prospects in Mycology and Plant Pathology*, International, Delhi, 1997.

Chaudhary, Sanjay: *Weed Management : Principles and Practices*, Narendra Pub, Delhi, 2011.

Clark, Vijay Paul: *Physiology of Crop Production*, International Book, Delhi, 2007.

Dabholkar, A.R. : *General Plant Breeding*, Concept, Delhi, 2006.

Dar, Ghulam Hassan : *Soil Microbiology and Biochemistry,* New India Publishing Agency, Delhi, 2010.

Das, Braja M. : *Advanced Soil Mechanics*, Taylor and Francis, Delhi, 2010.

Devi, C.R. Sudharmai : *Analytical Procedures in Soil Science and Agricultural Chemistry*, Agrotech, Delhi, 2004.

Erik Nissen-Peterson : *Rainwater Catchment Systems*, UK: Intermediate Technology Publications, 1999.

Ghosh, N.C. and K.D. Sharma: *Groundwater Modelling and Management*, Capital Pub, Delhi, 2006.

Gour, H.N.: *Integrated Plant Pathology*, Scientific, Delhi, 2009.

Groves, R H : *Weed Risk Assessment*, SBS Pub, Delhi, 2009.

Gupta, O.P. : *Modern Weed Management : With Special Reference to Agriculture in the Tropics and Subtropics*, Agrobios, Delhi, 2011.

Gupta, P.K. : *A Handbook of Soil Fertilizer and Manure*, Agrobios, Delhi, 2011.

Gustafson, A.F. : *Conservation of the Soil*, Biotech Books, Delhi, 2011.

Helen Bannayan: *Water Resources of Jordan: Present Status and Future Potentials*, Amman, Friedrich-Ebert-Stiftung, 1993.

Herminie Broedel Kitchen: *Soils and Crops : Diagnostic Techniques*, Satish Serial Publishing, Allahabad, 2004.

Husain, Ahmad : *Environment and Water Resource Management*, Sumit Enterprises, Delhi, 2006.

Jackson, E. : *Crop Management and Soil Conservation*, Biotech Books, Delhi, 2011.

Jana B L : *Water Harvesting and Watershed Management*, Agrotech, Delhi, 2008.

Joshi, N.C. : *Manual of Weed Control*, Researchco, Delhi, 2001.

Kanmony, J. Cyril : *Drinking Water Management : Problems and Prospects*, Mittal Pub, Delhi, 2010.

Kapoor, R.L. and M.L. Saini: *Plant Breeding and Crop Improvement*, CBS, Delhi, 1997.

Kataria, T N : *Plant and Crop Physiology*, Pearl Books, Delhi, 2008.

Khan, Samiullah: *Plant Breeding Advances and in vitro Culture*, CBS, Delhi, 1997.

Kumar Shailesh : *Plant Tissue Culture—Theory and Techniques*, Scientific, Delhi, 2009.

Kumar, M. Reddi: *Biological Control of Plant Pathogens, Weeds and Phytoparasitic Nematodes*, B.S. Pub, Delhi, 2008.

Kumar, Pushpam : *Economics of Soil Erosion : Issues and Imperatives from India*, Concept, Delhi, 2004.

Linda M. Welkom: *Groundwater Chemicals Desk Reference*, Chelsea, MI, 1990.

Majid, F.Z. : *Aquatic Weeds : Utility and Development*, Agrobios, Delhi, 2010.

Majumdar, S.P. and R.A. Singh: *Analysis of Soil Physical Properties*, Agrobios, Delhi, 2008.

Maliwal, G.L. and K.P. Patel: *Heavy Metals in Soils and Plants*, Agrotech Pub, Delhi, 2011.

Mandal, R.C. : *Weed Weedicide and Weed Control Principles and Practices*, Agrobios, Delhi, 2010.

Mathur, S.M. : *Aquatic Weeds : Problems, Control and Management*, Himanshu, Delhi, 2005.

Millar, C.E. & L.M. Turk: *Fundamentals of Soil Science*, Biotech, Delhi, 2001.

Mohanty, A.K. : *Entrepreneurship in Agriculture : Scopes and Opportunities*, Agrotech Pub, Delhi, 2011.

Musil, Albina F: *Identification of Crop and Weed Seeds*, Scientific, Delhi, 2003.

Narasaiah, M. Lakshmi : *Energy, Irrigation and Water Supply*, Discovery, Delhi, 2004.

Pareek O. P.: *Advances in Horticulture: Fruit Crops,* New Delhi, Malhotra Publishing House, 1993.

Patel, GM: *Botanical Pesticides for Pest Management*, Scientific, Delhi, 2008.

Prabakaran, G. : *Introduction to Soil and Agricultural Microbiology*, Himalaya, Delhi, 2004.

Raju, R.A. : *Field Manual for Weed Ecology and Herbicidal Research*, Agrotech, Delhi, 1997.

Rob Jenkins and C.K. Jain: *Advances in Soil-Borne Plant Diseases*, Oxford Book Company, Delhi, 2010.

Sahoo, Dinabandhu: *Farming the Ocean : Seaweeds Cultivation and Utilization*, Aravali, Delhi, 2000.

Sankaran, K.V. : *Alien Weeds in Moist Tropical Zones : Banes and Benefits*, Kerala Forest Research Institute, Delhi, 2001.

Saraswat, V.N. : *Weed Management*, Indian Council of Agri Res, Delhi, 2003.

Sathyanarayana B N : *Plant Tissue Culture : Practices and New Experimental Protocols*, I K International, Delhi, 2007.

Seema Srivastava: *Plant Physiology and Biochemistry*, Campus Books, Delhi, 2009.

Shanmugavelu, K.G. : *Weed Management of Horticultural Crops*, Agrobios, Delhi, 2004.

Sharieff, Afzal : *Fundamentals of Soil Geography*, Sarup Book Pub, Delhi, 2010.

Sharma, Vijay Paul: *Glimpses of Indian Agriculture : Macro and Micro Aspects*, Academic Foundation, Delhi, 2008.

Singh, Satish Kumar: *Fundamentals and Management of Soil Quality*, Westville Pub, Delhi, 2009.

Singh, Surjit and U.S. Walia: *Identification of Weeds and Their Control Measures*, Scientific Pub, Delhi, 2010.

Singh, Yadvinder and Bijay Singh: *Green Manuring for Soil Productivity Improvement*, Daya, Delhi, 2007.

Somani, L L and P C Kanthaliya: *Soils and Fertilisers at a Glance*, Agrotech, Delhi, 2004.

Sudharmai Devi: *Analytical Procedures in Soil Science and Agricultural Chemistry*, Agrotech, Delhi, 2004.

Swarup, Ram : *Elements of the Nature and Prospectus of Soil*, Manglam Pub, Delhi, 2011.

Thakur, Chandrika: *Scientific Weed Management*, Syndicate Pub, Delhi, 1993.

Trivedi, P C : *Plant Physiology : Current Trends*, Pointer, Delhi, 2007.

Tyagi, I.D. : *Plant Breeding and Genetics at a Glance*, South Asian, Delhi, 2005.

Verma, L.R. and R.C. Sharma: *Diseases of Horticultural Crops: Fruits*, Indus, Delhi, 1999.

Index